Annie's all about Goats
Essential Goat Care

By Mary L. Humphrey

Other Books by Mary

Essential Lotion Making, *Skin Care Made Easy*

Essential Soap Making *(Mary Humphrey and Alyssa Middleton)*

Advanced Soap Making, *Removing the Mystery (Mary Humphrey and Alyssa Middleton)*

His Pasture Press

Annie's all about Goats: Essential Goat Care

Editing by Crowsourced Team: Starla Ledbetter, Sharon Parker, and Debbie Richards.

Cover by: Jennifer Smith, Owner-Operator at Eco-Office Gals.

ISBN-13: 978-0692480182 (His Pasture Press)
ISBN-10: 0692480188

Dedication

I dedicate this book to all of the people in the world that love nature and adore animals. Your eyes hold real peace, and your heart is filled with caring.

Until one has loved an animal, a part of one's soul remains unawakened. – Anatole France

Some people talk to animals. Not many listen though. That's the problem. – A.A. Milne

Mary L Humphrey

Table of Contents

Mary L Humphrey

Acknowledgments

Over the span of a decade, I have received uncounted requests from friends, neighbors, and strangers asking for advice on goat care. I have often jumped in my truck to help with last minute birthing issues, or health and wellness care. I thank God for these opportunities, as well as for the people that have come into my life through calls for help, and who have joined me in my barn for mini-classes and instruction. Each has given me the opportunity to share and promote the wonderful world of goats.

I am grateful for my husband, my coach, my lifetime partner, the one that supports me as a person and makes sure I follow my endeavors, passions, and God-given gifts.

Thank you to my friend and coach, Debbie Schnitzius, who encouraged momentum and completion of this book. When I held onto the side of the cliff of frustration and wanted to let go, she made sure I hung on and got back on track. Debbie guided me to follow my God-sized dreams.

I am thankful for my editing team. Each invested their own share of time that pulled their attention away from their own businesses, personal, and farm hours to edit this book. Thank you Starla Ledbetter, Sharon Parker, and Debbie Richards.

Above all, I am thankful to God for the life and joy that he has given me. One of my constant goals in life is to help and encourage others, and this book has given me the opportunity to do so.

Chapter 1
Things to Consider
Before Buying Goats

When our barn began filling with goats I quickly learned how very different a goat was to care for compared to any other animal I had owned.

Most of us are familiar with raising cats and dogs. We dive into caring for them from weaning forward by following simple guidelines of good nutrition, clean water, exercise, and routine vet care. Goats, however, even those we consider as pets, are livestock that require additional expertise, time and management. Goats are definitely not plug and play animals. People who have never raised a goat often categorize them as "grass mowers", but their needs go well beyond that activity.

When purchasing a goat there are numerous things to consider, such as parasite prevention, fencing, feed, water, housing, breeding, disease, and hoof care.

Successful goat management is not an exact science due to the many factors involved. All farms are not alike, taking into consideration different soil, trees, pasture, and predators. Goat breeds also have distinct characteristics. One breed may be more aggressive, another may be more vocal. Some breeds are more inclined to jump the fence than others.

Genetics vary amongst goats as well. Some goats within the same breed, depending upon inherited genes, may be more or less hardy, or perhaps prone to birthing issues, and may even have stronger personality traits.

Before purchasing goats:

- Know who you are purchasing from
- Plan, and already have in place prior to the purchase, fencing and housing
- Determine feed costs and availability
- Plan for breeding (or not to breed)
- Study basic goat care
- Purchase more than one goat (at least two)
- Find a goat-knowledgeable vet
- Prepare a goat care medical kit
- Check pasture for poisonous plants
- Invest in a livestock guard animal
- Prepare for issues (health, fencing, birthing, etc.); there will be issues!
- Check local livestock zoning regulations
- Determine the amount of time you will be able to devote to goat care
- Know that livestock care involves 365 days of the year, and irregular hours that pop up on some days

Chapter 2
Purchasing Goats

Purchasing goats is not complicated; it can be an enjoyable experience, however, the future of your herd depends greatly upon the initial goats that you bring to your farm.

Buy More than One

My very first piece of advice for prospective goat owners is to purchase more than one goat. Goats are social animals. They thrive when they share activities with another goat or two. A lonely goat can quickly become a sick goat. At the very worst, a goat may grieve itself into illness and death, and at the very least, it may become very vocal, seeking attention, leaving you - the herdsman - in dire need of peace and quiet.

Start your herd out small, and grow it gradually. Give yourself time to experience the ins and outs of goat care. If you plan to breed goats, breed just a small number the first season. One goat is too few on a farm, but you can easily start out with too many and quickly become overwhelmed.

Know Your Purpose for Purchasing Goats

Are you purchasing goats for pets, or for breeding? Is your goal milk, fiber, or meat? Are you purchasing for show, and are you seeking pedigrees or registered goats? Never purchase goats solely because they are cute. Hobby goats are perfectly enjoyable, but make educated decisions based on your (and their) needs.

Know the Goat Breeds

Research each goat breed. When I first entered the world of goats, I quickly noticed that most goat owners have favorite breeds, especially the breeds they currently own. Do not feel pressured into acquiring a breed of goat you are not interested in owning. I devoted Chapter 4 of this book entirely to goat breeds because each breed has their own unique set of traits, and they serve distinct purposes – dairy, meat, and fiber.

Know the Breeder and Their Farm

Personally knowing the breeder before you make a purchase is not always possible. At the very least, visit the farm more than once if it helps you to feel comfortable. Observe the goat housing, pasture, and pens, are they clean and well kept? Talk with the breeder in length. An honest goat farmer will tell you everything you need to know, nearly beyond what you want to know. They should be knowledgeable about the maintenance of their own herd, and happy to discuss details with you.

Know the Goat's Health

Observe the herd. Do the animals appear healthy and well cared for? Ask the owner what health problems they have experienced, if any. What wellness or medical care have the goats received?

If you want to purchase Caseous Lymphadenitis (CL) or Caprine Arthritis Encephalitis (CAE) free animals (see pages 15 and 77), ask the herd owner if he has had goats with either disease in the herd, and ask for proof of testing.

Before visiting the farm, write down the questions you want to ask and have extra paper on hand to take notes. Each note you record is the start to the health records you keep for any goats that you purchase. If you are like I am, written notes serve better than memory.

Awareness of the health of each goat is the main thing I stress to people before they plop their money down. Think about how you purchase a used car. What if you did not listen to the engine running, test the brakes, or look under the vehicle for corrosion before you signed on the dotted line and drove the car off the sales lot? You might sink your money into a purchase that becomes an enormous money pit, and in comparison to purchasing goats, could even lead to heartbreak when you end up with a sick or dead animal.

You will also want to observe if the goats are dehorned, vaccinated, dewormed, and if their hooves have been properly trimmed.

Speaking of horns, if the goats you are observing have no horns, ask if the goats have been disbudded (if their horns have been removed), or if they are naturally hornless (polled).

What is the Age of the Goat?

Ask for the age of the goat. You do not want to purchase an older goat if you plan to breed for many years. On the other hand, I have purchased slightly older goats, between 3 – 5 years old, from production herds. They were healthy does that birthed wonderful kids and later retired on my farm. Always keep in mind, mature goats make great pets!

The approximate age of a goat can be determined by looking at their teeth (see Teeth and Age, page 59).

Is the Doe Bred?
Is the doe bred? Is she old enough to be bred? If she is not old enough for breeding, was she housed separate from the bucks? When purchasing a full-grown doe, has she had any problems with prior births?

If the goats are bred, the owner should be able to tell you when the breeding occurred. Write the breeding and expected due dates down for your own future reference.

If you purchase bred registered does, ensure you receive completed documentation from the breeder before you leave the farm. In some cases, the breeder may need to complete paperwork after the kids are born. In either of these scenarios, you should walk away from the purchase with no less than copies of the documents, along with the pedigrees.

Know their Feed
Ask what type of hay and grain is being fed, and how often and how much. To avoid creating additional stress, and to prevent digestive upset, consider buying an initial supply of feed and hay from the seller, or purchase the same feeds from a farm store. This will allow you to change the feed gradually if you decide to do so.

Temperament
Do the goats appear to be afraid of people? Are they overly jumpy? Do they exhibit pronounced behaviors, such as extreme jittery movements, which could

indicate a nervous condition, or could be a sign of mistreatment? If you want tame goats on your own farm, you do not want to purchase those that are on the opposite end of the calmness scale.

Local Livestock Transportation Regulations

Know your local livestock transportation regulations. Some states require the tagging or tattooing of livestock before the sale, and some require documentation and record keeping.

Federal laws exist in the United States that specifically pertain to hauling livestock across state lines. Ask your local veterinarian or livestock agriculture extension office, or department of agriculture, for detailed instructions.

Our First Goat Purchases

When we first purchased goats, we bought what was pleasing to our eyes. Twin bucks that were "cute," smaller does because we did not want to "take the best ones" from the breeder, and the list of bad choices went on from there. The beginning of our farm was fun, but we ended up with too many animals that did not fit into our farm vision, which led to the reality of nearly having to start over within our first year. We quickly learned to make better choices!

Mary L Humphrey

Chapter 3
Bringing New Goats to Your Farm

When introducing new goats to your farm, whether you are bringing in goats for the first time, or if you are presenting new goats to an existing herd, there are simple, but important, measures to follow.

Stress

Goats have a tendency to stress when moved to a new location. Be prepared for an excited goat. Ensure the goat cannot break loose while being handled. Use a harness or collar if possible, and be careful to have a firm grip with your hands before opening a gate or trailer door. Goats are experts at twisting and wiggling away. Do not let size, breed, or an appearance of calmness fool you. Be prepared for their plan of action, not yours!

Pecking Order

Goats will settle into a pecking order within the herd. Levels of dominance always exist. Goats challenge each other through their eyes, body language, and fighting (head butting). Do not become alarmed by aggressive behavior however, you may want to separate the younger and much smaller goats if extremely rough behavior erupts and continues.

When introducing new goats to your herd, observe them through several feedings to ensure none are starved out through intimidation.

Horns or no horns, we have not experienced serious injuries resulting from head butting. What you do want to be aware of, however, goats with longer horns can lock their horn under another goat's leg, which can eventually cause a break or sprain. If serious fighting ensues, separate the rivals until they settle down. It might take several days for the goats to calm down, but then, I have experienced a goat or two that never did get along. I ended up keeping those particular animals in separate areas.

Feed

Knowing what the prior owner fed is a plus, and a must, when bringing new goats to your farm.

Goats stress and may not want to eat when they are in new surroundings. In these cases, it helps to feed what the prior owner was offering --- familiar grain and hay. Gradually switch to the grain and hay that you will be feeding on your farm. Doing so helps to prevent digestive upset.

Illness Prevention

Each farm carries its own unique bacteria and parasites. The goats on your farm may be immune to whatever is in your soil, but new goats may not be. The opposite is also true, when introducing new goats to your herd, quarantining them helps to ensure disease does not spread from the new goats to the others. Consider an incubation period of two to three days to observe the behavior and health of the new goats.

Our Early Experience with Transporting Goats

We had just purchased a dozen beautiful Spanish cross does. We pulled the trailer of new goats up to the gate. Just as we opened the trailer door, I turned my head. Out popped a young doe. I chased her through fields of waist high weeds. I never gave up, but I was convinced we had just lost one of the animals that we so carefully selected. Luckily, after an hour of following the determined doe, I stumbled across an old fence line and became smarter than a goat. I tricked her into a corner. Never again did I let my attention roam when handling a new goat.

Our Experience with Moving Goats to a New Farm

Sometimes we need a course in goat psychology...but then, if we get to the point where we feel we understand them today, we might not understand them tomorrow. It seems we always have something new to learn!

We moved a small handful of our herd to our new farm.

We did not mow the tall grass and vegetation in the new pasture before the move. We knew the cedars and the tree saplings would be nutrition-filled delicacies.

We chose the thinnest, oldest, and most "needy" of our goats for the first move. I wanted them to experience the joy of the fresh new pasture.

Surprise!

We should have known. In fact, we did know, but never thought it would happen to our goats. The newly located goats did not touch the greenery for the first 3 days. Each day when we arrived to feed supplemental hay, which we decided to do just in case, we noticed the does were losing weight.

We decided that they were, 1) stressed because we moved them, 2) spoiled to their old barn and twice-a-day hay feedings, and 3) they missed the rest of the herd.

Lesson learned.

Goats are creatures of habit!

Chapter 4
Goat Breeds

There are roughly 200 goat breeds around the world, and that number continues to change due to the discovery of new breeds in isolated places. I describe the most common domestic goat breeds in this chapter.

Goat breeds vary by size, shape of ears, color, and function; which includes dairy, meat, and fiber producers.

To make an educated decision on the type of goat that fits your needs, review the basic qualities of each of the breeds, talk with numerous goat owners, and observe the breeds that interest you in-person.

Dairy Breeds:
Dairy goats are known for their milking ability.

In my herd, the highest volume of milk was produced by the Saanan breed, followed by the Alpine, and lastly by the Nubian.

Nubian milk in my herd was by far the highest in butterfat, with the Alpine breed next in comparison.

Keep in mind, genetics play into any trait, meaning, volume and butterfat characteristics are passed down through the genes. Milk volume and content are also highly subject to nutrition and the overall health of the animal.

Alpine
A medium sized breed, with erect ears.
- Easy to train
- Large volume of milk
- Loyal
- Multiple coat colors
- Hardy
- Straight Face
- Coat colors vary but should not be solid white or light brown with white markings.
- Seasonal breeders

LaMancha
Small to medium sized breed with very small ears, often appearing as if they have no ears
- Straight nose
- Calm and docile
- Productive in milk
- Intelligent
- Any color coat
- Seasonal breeders

Nigerian Dwarf
Miniature breed with medium length erect ears
- Rich milk - high butterfat
- Straight or slightly dished face
- Occasional blue eyes
- Any coat color or pattern
- Breed year round

Nubian
One of the larger breeds, with pendulous ears that extend about 1" below the muzzle
- Vocal
- Easy to train
- Rich milk – high in butterfat and protein
- Hardy
- Gentle
- Less milk than other dairy breeds (adequate volume)
- Convex (Roman) nose
- Multiple coat colors
- Seasonal breeders

Oberhasli
Medium to small sized breed with erect ears
- Straight nose (must not be Roman)
- Bay (chamoise) coat color (female coat can be solid black in color)
- Black dorsal stripe
- Black udder, belly and below the knees
- Nearly black head
- Moderate milk production
- Seasonal breeders

Saanan
Largest of the dairy breed, with erect ears
- Easy to train
- Large volume of milk
- A good balance of milk properties – butterfat and protein
- Quiet
- Hardy

- Pure white or very light cream coat
- Strait or dished face
- A breed commonly used around the globe by dairies
- Seasonal breeders

Sable
Colored Saanan, having any color combination except for pure white or very light cream

Toggenburg
Medium sized breed with erect ears
- Light fawn to dark chocolate coat
- White ears and white lower legs
- Side of tail and two stripes down face are white
- Shaggier coat than other dairy breeds
- Not as docile as other dairy breeds
- Average milk production
- Seasonal breeders

Meat Breeds:
These types of goats originated as breeds that supplied meat, called chevron or mutton.

Meat goats grow rapidly, and they quickly bulk up in weight as they mature.

Boer
- White coat with dark head (occasional solid dark or spotted coat)
- Pendulous ears
- Large breed
- Gentle
- Strong
- Hardy

- Horns curve backwards
- Heavily muscled
- Rapidly bulks up
- Breed year round (Boer goats lean toward seasonal breeding in areas that have shorter periods of daylight in the winter months.)

Kiko

- Hardy – very resilient, requiring less management
- Rapidly bulks up
- Large framed
- White coat – occasionally colored
- Coat characteristics depend upon climate and geographic location – sleek in summer, longer coat in winter
- Breed year round
- Excellent mothering and milking ability

Myotonic

- Also called Tennessee Fainting Goat
- Small to medium size (not miniature breed)
- Several different ear types, but must have crimp down the middle
- Long or short coats
- Variety of coat colors
- Genetic characteristic - myotonia congenital gene – causes leg muscles to lock up when startled, hence, the goat loses balance and falls over (appearing as if it has fainted)
- Breed year round

Savanna
- Hardy
- Excellent grazers
- Fast growth
- Strong legs and bones
- Muscular
- White to cream colored coat
- Black pigmented skin – protection from ultra-violet rays
- Breed year round
- Excellent mothering and milking ability

Spanish Meat Goat
- Medium sized – short, stocky
- Any color - multi color and spots are common
- Hardy
- Heavily muscled
- Breed year round

Fiber Breeds:
Fiber goats are any breed that produces cashmere wool from their coat, or mohair, which is the soft inner layer of the coat.

Angora
- Curly coat – approximately 6 to 7" long
- Adult produces an average of 12 lbs. of fiber per year
- Docile
- Not known for hardiness (research the breed – check your area and climate for hardiness)
- Seasonal breeders

Other Breeds:

There are many breeds not listed in this book, most of which are less common. I end my list of breeds, however, with the Pygmy goat. This breed is classified as a meat goat, but they are typically raised as pets or show goats due to their small size.

Pygmy
- Small breed
- Considered meat goats but generally are raised as pets
- Heavy boned
- Active
- Hardy
- Variable colored coats with specific markings for those that are registered (for show purposes)
- Breed year round

Our Personal Thoughts on Goat Breeds

As you can see, there is a wide variety of goat breeds available. To add to the mix, personality and genetics differ from herd to herd. To make an educated decision on the breed of goat that fits your needs, review the basic qualities of the each of the breeds, read as many materials as you can, talk with goat breeders, and observe the breeds for yourself.

Now that you have read factual breed information, I will share my personal favorites with you:

Nubian – Sweet personality and gentle. Vocal. Nubians definitely voice themselves more than the other breeds. Their milk is rich with butterfat.

Snubian – Cross breed produced from full-blood Nubian buck and full-blood Saanan doe. 1st generation: solid white coat. 2nd generation (bred full-blood Nubian buck to Snubian doe): solid white coat, or cream to mottled cream coloring. Sweet and loyal personality. Eager to milk, with the ability to produce proliferous volumes, both on the stand and while nursing kids. Volume vs. butterfat balance excellent. Large and fast growing, hardy kids.

Boer Nubian – Cross breed between Boer buck and Nubian doe. Produces kids with hardy Boer growth characteristics. Excellent butterfat rich milk. Beautiful coat colors and markings.

Note – In my herd, I cross bred to bring additional hardiness into the genetics. Cross breeding produces fast growing and resilient kids.

Chapter 5
Fencing, Housing, Storage

Fencing

You may have heard the age-old saying, "If you can see through a fence it is not goat proof." Well, that is right on target! One of the most challenging aspects of raising goats is to keep them on the right side of the fence, inside the boundaries. They are most definitely proliferous escape artists.

Fencing must be tall enough to keep the goats from jumping over. Have you seen a deer gracefully glide over the top of a fence? Picture that, and welcome to the world of goats.

Fence height should always be at least 4 feet, and may need to be taller depending upon the size of the goats, or the predators that are local to your area.

Rigid metal wire fence panels are ideal for small paddocks, or for pasture fencing if you can absorb the cost. Secure panels to fence posts with metal clips or common fencing wire. Panels are easy to cut to size, and it is helpful to have a couple of extra pieces on hand for quick fence repairs, or to set up pens for birthing, confinement, or even an impromptu hay feeder. Another plus to rigid wire panels; they are easy for one person to handle, especially for repairing fences.

Ensure the squares in any type of fencing you use for goats are no larger than 4"x4". Goats are notorious at getting their heads stuck in fencing, with or without horns. When constructing smaller areas, such as barn stalls or pens, you might want to consider rigid wire panels manufactured for other types of livestock, such as pigs. These panels have much smaller spaces along the bottom rows, which prevent kid goats from slipping through. Yes, they will purposely do that, sometimes when they are just a few hours old.

Goats enjoy standing with their front hooves up on fences, especially when they see their owner approaching at feeding time. A fence not tightly stretched when installed will eventually sag, which can also result in gaps under the fence, which makes easy access for predators and young kids.

Electric fencing is another option, consider running a hot wire at ground or top level (or both), as an added bonus to woven wire or solid panel fencing. Goats will step through electric fence wire that is widely spaced.

Gates are also an important factor to consider when building fences. I prefer gates made from sturdy metal tubing. Make sure the gate is tall enough to prevent goats from jumping over, and low enough to the ground to prevent small animals from crawling under. When we have young kids on the farm, we tie small sections of rigid wire panels to the lower sections of the gates.

Consider having your fence installed by a professional. If you do so, educate yourself beforehand so that you will be able to tell the installer exactly what your needs are. To save on costs, consider purchasing the fencing supplies yourself, which leaves you with only the expense of paying the installer for labor. Keep in mind, there are specific steps involved with installing a woven wire fence that a professional, or a person very knowledgeable with livestock keeping, can assist you with, such as stretching, installing the correct number of clips at the metal fence posts to prevent sagging – at least 5 or 6, and cross bracing.

Visit farm stores and talk with personnel, most are well versed in farm and livestock care. They will guide you through all of your fencing questions and needs. You may also contact your county livestock extension office for a listing of businesses that install fencing, and the local officials may be able to provide you with additional fencing and native-to-your-area predator advice.

Pasture and Paddocks

Ideally, install fencing on flat ground. If you do not have flat ground, avoid the placement of fencing on the low side of a slope or hill. Goats eventually take notice of low spots and they will jump over.

When planning the layout of your fencing, consider your farm needs, such as kidding, breeding, separating does from bucks, milking, isolation pens, aisles and alleys for feeding, cleaning or tractor movement.

To calculate the number of goats that an acre can sustain, know that this process is different than it is with other types of livestock. With goats, the focus is not on how much green pasture is available, although that is part of the overall goal. It is more important to concentrate on the prevention of parasite overload. In general, poor land may support 2-4 goats per acre, while lush land may support 5-10 goats per acre.

Goats eat plants that other animals will not touch, for example: poison ivy, and because they have a very high metabolism rate, they must eat frequently.

Goats generally graze from the top of the plant down, unlike other livestock, so it is best to rotate pasture to ensure they are eating above ground level.

When planning your farm, it is important to consider the rotation of pasture to prevent parasitic overload. Separate your pasture into sections, called paddocks, and move the goats from pasture to pasture in approximately 90-day intervals to ensure parasite larvae die off. During warmer seasons, you may have to mow or bale the unused pasture. If grazing drops the pasture down below 4" in height, the chance of parasite infestation increases.

Select dry land for pasture. It can contain a stream or water source, but the land must have good drainage. During critical times, very damp seasons when the ground becomes saturated, avoid disease and parasites by feeding the goats baled hay in a dry lot. Parasites tend to proliferate in damp areas and during extremely humid and rainy seasons.

Tethering
A goat's main defense to predators is their ability to run and scatter. If you must tether (tie the goat to an object), only do so when you are outdoors with the animal. Never leave a goat on a chain or rope permanently.

Our Experience with Herding Escaped Goats
Some goats are very difficult to corral when they are not confined. When attempting to catch one, do not look the goat square in the eye. A goat's pupil is horizontal. This allows them peripheral vision that is much better than ours is, nearly as if they see around corners. When you stare at a goat they will stare back, and they are very aware of your movements as you try to capture them. The best method is to appear as if

your attention is on another goat or object, not on the goat itself. You have to fool them to catch them! Never make any quick movements towards a goat until you are within close range, and then be prepared to grab a leg or hoof, without injuring them of course, and be ready to hang on tight if the goat is not tame.

Now that I have told you what not to do, I will tell you the best way to catch goat is through feed. Act as natural as possible, and place the feed as close in front of the goat as you can get. More often than not, the goat will entertain itself with the feed, allowing you to move in and make the catch. Better yet, you may try carrying the feed through an open gate, or through the broken area of the fence line (where the escape may have happened). Goats will often follow you back to a feeding or secure area, because they, too, want to feel comfortable and safe.

One of Our Experiences with Escapee Goats
Every day a small group of our young does trekked their way from inside our fenced in goat pasture down to the lower open field. We called the little troopers our "escapees." Their method of break out was not easy to determine.

Goats can be tricky. If they catch on to being watched they may not act natural. Turn your back, when they know you are not watching, they will do exactly what you were trying to spot them doing in the first place.

In pursuit of determining how our escapees were getting out, after walking the fence line and not finding a fence break, my husband and I laid down a plan of action. He stood in an upper area where the goats could not see him. I stood behind a tree just below the suspected escape spot. Our undercover surveillance team succeeded in just a manner of minutes.

The first goat stuck her head through a tiny break in the fence. As she pushed her body forward, the hole expanded. After four goats swiftly passed through the same location, the hole popped back as if there were no break, none at all!

With wire and pliers in hand, we fixed that problem. Always be prepared for a quick fence repair!

Goat Housing

Goat housing does not have to be picturesque or elaborate; it just needs to provide adequate shelter and protection from the elements.

Basic housing requirements:
- Shelter from rain (snow, sleet, etc.)
- Shelter from chilling winds
- Bedding, and an area that allows for easy and frequent floor cleaning (raking/scraping)
- Areas to separate the goats (bucks from the does, maternity pens, newborn kids, and sick animals)
- Access to water
- Ventilation (do not confine goats without a window, panel, or door to open for fresh air)
- Lighting (optional, but does help with birthing, milking, and when tending to sick animals)

Hoop houses work well for breeding, or for small paddocks. Bend a long rigid wire panel to form an arch and tie on a heavy tarp to make the shelter waterproof. Use stacked bales of straw for an added wind barrier or insulation.

You may also consider building a simple lean to (two sided) on the side of an existing barn, or consider building a three sided shelter within a paddock. Ensure the open side of the shelter faces south for warmth, and have a plan in place to make the shelter more air and moisture tight if inclement weather makes an appearance, especially if you plan to breed or raise young kids.

When building a barn, or any building that houses livestock, **do not use treated lumber**. Treated lumber contains chemicals that are toxic to animals, which is especially dangerous to livestock that likes to nibble – goats!

Our Experiences with Barn Spoiled Goats

Our experience is that goats do not like to get their hooves wet. On a wet day, we appreciate a loafing area in the barn where the goats wait out the rain. Our neighbors have said, "Your animals are barn spoiled." I think we can safely say that all goats could be happily barn spoiled. If they have no place to escape from a rainstorm, they will bunker down in the tiniest of places to keep themselves out of the moisture.

Electricity and Lighting

Access to electricity and lighting in the barn is ideal if you intend to breed, milk, or show your goats. You can easily do all of these things without electricity, but you will be dealing with darkness, with no access to power equipment such as heat lamps, electric clippers, or a milking machine.

Fiberglass panels installed by the piece in a barn roof, in replacement of lengths of sheet metal, allow light to filter through.

If you have electricity and lighting in your barn, and if you plan to breed goats, you may consider a barn camera (barn cam), however, these are definitely not a necessity. Through a cable wire, or through a wireless signal, a barn cam sends video images to a television or computer monitor inside your home, or to your cell phone. Less expensive options, if your barn is within close proximity, are battery operated audio monitoring devices commonly used within a home for infants. You may consider a plug in intercom system (left in the "on" position while listening). Again, listening or watching the barn is not an essential requirement.

Bedding and Flooring:

Dry and clean bedding is extremely important for disease prevention as it absorbs urine and helps to disburse feces.

Bedding must be free of mold to prevent illness. Wheat straw is my preference for bedding, however, there are other suitable types available: uneaten hay, dried pine or cedar shavings (not regular wood chips, which may promote mastitis and other diseases), and bagged pellets that are available at farm supply stores.

Even though clean and dry bedding is encouraged, the deep litter method provides warmth in colder climates during the winter months. A dry sand or dirt floor, covered with several inches of litter (used bedding), provides insulation – a comfortable place for the goats to sleep and rest. Rake the top layer of litter off as it becomes soiled, and replace with fresh bedding.

Concrete floors tend to remain cold and damp in inclement weather. This is why I prefer gravel or dirt floors in a barn. A layer of unwashed gravel, which contains small particles, will pack down and become a hard surface, which promotes the drainage of urine. Spread bedding, such as straw, on top of gravel in the loafing and sleeping areas of the barn.

Exposed gravel in other areas helps to prevent the overgrowth of hooves.

Another barn flooring system that drains well is a 6" layer of gravel on top of a dirt floor, covered with landscape fabric, and topped with 6" of sand. Urine filters through the sand and gravel, down to the dirt floor. This type of bedding is beneficial for pulling the dampness and ammonia odor away from bedding. To avoid sand sticking to a newborn kid, I do not recommend sand bedding in birthing pens.

Use lime to freshen dirt floors. Lime also helps control the fly population in humid and warm weather. After raking or scraping soiled bedding from the barn stalls, which exposes the dirt floor, sprinkle a thin layer of lime on the soil. Allow the floor to air dry as long as possible before spreading fresh bedding. Keep in mind, there are several types of lime available, **do not use the hydrated type!** Bags containing hydrated lime are labeled with caustic warnings. This type of lime can cause injury to both you and your livestock. Agricultural lime, or ag – garden – or barn lime, or dolomite, is gray in color, and is considered safe for livestock. This type of lime serves as an antibacterial treatment to the floor. It helps to dry urine, and rids the floor of the odor of ammonia. When purchasing lime at the feed or farm store, be very persistent and cautious, again, and make sure you are purchasing agricultural lime, **not hydrated lime.**

If you can afford the additional cost, there are name-brand granular and pelleted products on the market that contain minerals and clays that control ammonia and help to keep stalls dry. Look for these products in the equine section of a farm store, or ask your local dealer for availability.

You may also consider purchasing sodium bicarbonate (commonly known as baking soda) in bulk quantities to sprinkle on and sweeten up a dirt floor before adding a layer of clean bedding.

Feed Storage

The proper storage of feed is of extreme importance to the health of your goats.

Feed (grain and hay) must be stored in a dry area. Dampness promotes the growth of mold and spoilage, which can lead to disease and death.

To prevent disease, feed must not be exposed to rodents. Do not allow cats to nest and defecate in the feed and hay storage areas. Exposure to either of these contaminations can lead to diseases, such as Toxoplasmosis, and abortions (see Chapter 8).

Goats are intelligent and often know your routines. They will not only try to open gates that you have passed through, but they will also try to get into feed rooms and storage areas once you leave the premises. Store your feed and hay in a separate paddock, room or shed, and ensure the door or gate is well secured, padlocked if you prefer.

Chapter 6
Livestock Guard Animals

A livestock guard animal should be a major consideration for the protection of your herd.

Livestock guard animals include dogs, donkeys, mules, and llamas. Other species of animals are often used, but they provide less protection. In fact, some guard animals are used as bait to draw predators away from the goats. I am not fond of exposing an animal to any level of risk, so in this chapter I discuss the animals that can actually fend off an attack.

Predators

Predators include coyotes, wolves, fox, and dogs – both wild and domestic from neighboring farms, large feral cats, and other predatory wildlife common to your region.

Predators are always a danger, but the threat level may increase during particular seasons. Coyotes, for example, are more likely to search for food (prey on livestock) when they are feeding their young in the spring.

Farmers utilize different stimuli to deter predators, such as leaving an outside barn light on, or playing a radio, or the application of an odor repellant – which, by the way, normally carries a short life. Predators become accustomed to the stimuli, so use it randomly. If you leave a light on, switch back and forth between it and a radio. If you use an odor repellant, replace it often and use a variety of types.

Livestock Guardian Dog (LGD)

A livestock guardian dog (LGD) is my preference of protection. They see well at night and have very sharp hearing. Unlike other species of guard animals, their keen sense of smell warns them of a predator long before it approaches the herd.

An LGD becomes part of the herd. He or she will sleep and live with the goats. Their attention is always on the goats, ready to fend off predators every waking moment of their day, and they are on alert while they sleep (as if they have one eye open at all times).

When selecting an LGD you do not want a breed that herds or chases after livestock. Goats will run away and scatter, instead of huddling from a herding approach. For this reason, I recommend the Great Pyrenees, Anatolian, and Maremma breeds.

In addition to selecting a particular breed of dog for an LGD, it is beneficial to bring in a dog or puppy (between 8 and 12 weeks old) that is already acclimated to goats. They are much less likely to chase after, or have an adverse reaction to the goats.

The sex of an LGD is not a big issue. Consider neutering a male to eliminate any desire to mount a goat. A female may be slightly more protective of young goats due to her mothering instincts.

A single LGD per paddock is preferred. The dog will not have a companion to play or compete with so their attention is directed solely towards the goats.

LGD's generally slumber during the day, always in close proximity to the goats. They are vocal at night as they deter any threat (seen or unseen) that approaches the herd.

Breeders will assist you with LGD training advice, but remember these important points: he is part of the herd, he does not spend his time bonding with you or seeking your attention, he knows the word "no," he is on guard 24 hours a day - 7 days a week, he protects the herd without scaring the goats away. He is a goat! He does not bark, chase, herd or bite his goats. He is a working dog, and must be treated as such.

Donkey (Mule)
Pros:
- Dislike dogs and coyotes
- Same diet as goats
- Females and geldings are easiest to work with and are less aggressive
- Low maintenance
- Long life, 20-25 years

Cons:
- If unaccustomed to goats, may not like a goat, therefore, may injure kids, and may be aggressive during feeding
- Dislike dogs and will intimate an LGD if nearby
- Some will work hard, some will hardly work
- Will jump fencing to get to a nearby donkey
- May become aggressive and overprotective during kidding season
- May step on or injure newborn or young kids

Llama
Pros:

- Dislike dogs and coyotes
- Geldings and females are preferred
- Will run towards predators; kick and paw it to death
- 12-18 years life span
- Easy to handle and train
- Works better without another llama in the paddock, becomes part of the herd
- Can protect a large number of goats at one time
- Grass eater that does not commonly bloat

Cons:

- A mature llama, unaccustomed to goats, may not like a goat
- May spit at humans
- Castrated males may injure goats when trying to mount
- Veterinarians that know how to treat llamas may be difficult to find
- Initial investment is high

Ostrich, Pony (Horse), and Alpaca

These animals are used as guard animals but I prefer that they be the ones that are guarded. A guard animal must possess the instinct to run towards the predator, and have a very keen sense of smell and hearing. They must have the ability to run after a predator, not with the herd, at speeds up to 40 miles per hour (coyotes can run at this speed). They must have the sense to become part of the herd – not to be used as bait.

Mixed grazing, a pasture that holds both goats and horses, or cattle, helps to deter predators, but these animals should also be guarded by a donkey (mule) or preferably a livestock guardian dog. A dog possesses an inbred guarding ability, coupled with a sharp sense of smell and hearing.

My Experience with Training a Livestock Guardian Dog

I purchased an LGD that was a mix of the Great Pyrenees, Anatolian, and Maremma breeds. He was 9 weeks old and had been born on a goat farm.

Despite his daily exposure to goats (the dog housing was located inside the goat paddock), I had to train him to become the guard dog that he eventually became. He persistently whined at me for attention. I patted him on the head while I tended to the goats, but I did not stop and give him individual attention. I talked to him as I walked by, telling him that he was in charge of the herd, and pointing to the goats as I did so.

When kidding season arrived, I had to take drastic measures when I realized the youthful LGD was nipping at the young goat's legs. He was being playful, but an LGD should protect their herd, and not scare them. I built a large pen and housed the bossiest of dams and their young inside the pen, along with the young LGD. Needless to say, I did not do this part of the training. A dam will fiercely protect her young. The goats respected that the dog was inside their pen; however, they also did not allow him to touch their kids. This broke our LGD of his bad habits. I did not trust him before this training, and I advise the same for you. **Do not leave a dog that you do not know alone with young livestock.**

Feeding was also an interesting time. The goats wanted to eat what the LGD had in his bowl. It was difficult to watch, but I allowed the LGD to claim his own territory when he ate. With growling, but without lunging or biting, he let the goats know that it was his food, not theirs.

The most important LGD farm rules are: he must always be a part of the herd. He resides with the goats, he is alert when they are asleep, he is on duty constantly, and he is not man's best friend...he is the goat's best friend.

Chapter 7
Feed and Nutrition

A goat's diet affects the animal in the present, and is a determining factor for their future. Proper nutrition is a solid foundation for growth, breeding, birthing, milking, hooves, coat, and disease resistance.

Goats are herbivores and browsers, not grazers. They will eat tree bark, bushes and brush, and other available vegetation before they eat grass. They are very good at picking out the most nutritious portions of hay, shrubs, and trees to maintain their own rumen function.

Goats are finicky eaters, much unlike the common saying, "A goat will eat anything." For this reason, I advise goat owners to spend more money on nutrition to avoid waste. The cheapest source of feed is good pasture with high quality forages, however, goats also benefit from a balance of hay, minerals, and supplements, such as grain.

The Rumen
The rumen is the first of four compartments of a goat's stomach; followed by the reticulum, omasum, and the abomasum.

Goats are ruminants. Microorganisms in the rumen break down ingested feeds into carbohydrates, starches, fiber, and energy. The rumen is where most of the nutrients from feeds are absorbed, and where heat is produced that keeps the goat warm.

Goats chew their feed, soften it with saliva, swallow, and then the rumen breaks the feed down by fermentation, in movements called rumination. During this process, the goat will regurgitate the contents from the rumen, and will chew it further before swallowing once again. This is referred to as "chewing the cud." The goat also burps excess gas that develops in the rumen during the fermentation process. This is necessary to prevent illnesses, such as bloat.

The rumen is located on the left side of the goat. You may be able to see the rolling rumen movements, but you can definitely hear the rumination with your own ears if you sit close enough, or use a stethoscope. Place your head against the goat's left abdomen and listen to the rumination. It sounds much like a growling stomach, gassy, but do not be surprised when these sounds take on a different note. Each type of feed produces different digestive noises.

The rumen requires plentiful roughage and fiber to work properly, and it must be fully functional to maintain animal health and performance.

Once the feed particles are small enough, after the goat chews the cud and regurgitates several times, the remaining materials pass to the reticulum.

The Reticulum
The reticulum is the second stomach compartment, often called the honeycomb because of the pattern of the interior lining.

Feeds separate into smaller particles in the reticulum, with the largest particles held in the lining for further digestion and absorption. The smallest particles, and water, pass to the omasum.

The Omasum

Water and nutrients are absorbed in the omasum, the third stomach compartment, and then the remnants pass to the abomasum.

The Abomasum

The abomasum, the fourth and final stomach compartment, is considered the true stomach. It produces acids, and receives enzymes produced by the pancreas, all of which prepare any remaining proteins for absorption in the intestines.

The Intestines and Secretion

The small intestine receives the remaining feeds from the abomasum. Feed remnants mix with secretions from the pancreas and liver, and bile from the gall bladder. Nutrient absorption is optimal in the small intestine. All remnants then pass to the final digestive section, the large intestine. Water is absorbed, and any remaining materials secrete as feces from the rectum.

Goat feces are round and pelleted, sometimes called goat berries, and are somewhat dry in consistency. As you will read in Chapter 8, Health and Wellness Care, feces may indicate illness when it becomes clumpy, pasty, watery, or off color.

Breaking Down the Nutritional Needs

Nutrients fall into several categories:

Protein - Protein is the most important nutrient to a goat. It supports necessary bacterial growth in the rumen. Muscle growth and energy depend upon protein intake. Protein digests nearly entirely in the rumen.

Carbohydrates – Fiber, starches, and sugars are the source of carbohydrates, which supply energy. Carbohydrates digest in the rumen and the small intestine.

Fats - Fats serve as an energy source, but must be limited in a goat's diet. Fats digest poorly in the rumen. High levels of fat in feed reduces rumen function, which results in less nutrient absorption, and the goat ends up eating less.

Minerals - Goats require many minerals for body function and prime production. Minerals such as calcium and phosphorus are required at a higher level. Minerals such as iodine, iron, iodine, magnesium, potassium, selenium, sodium (salt), sulfur and zinc, are essential, but are needed at a lower level. Each of these lower level minerals is necessary to metabolize calcium and phosphorus, and is required for the development of muscle tone.

The ratio of calcium to phosphorus in a goat's diet is important to optimal health and must be about 2:1. Example, a cup of grain, or loose minerals containing 1.00% calcium, should also contain .50% (no less than .35%) phosphorus.

Vitamins – Vitamins are an important factor in the diet of a goat, however, some vitamins come from feed and some produce in the rumen. Vitamins that are fat soluble must be provided in the diet, such as vitamin A, D and E. Water soluble vitamins are created by the bacteria in the rumen, such as B complex. It is not necessary to include water soluble vitamins in feed, however, when the goat is ill, when the rumen is not functioning properly, supplementation must be considered.

Energy – The rumen processes some feeds into energy, such as fiber (hay, browse from trees, and pasture). Energy is vital, as it helps the goat to remain active and to retain warmth.

Water – Water is necessary to maintain life. The more water a goat consumes the more feed they intake. Water is essential to a goat's growth and milk production. Water must be kept fresh and clean, never frozen, nor in direct sunlight. Water directly exposed to harsh sunlight heats considerably, which results in algae growth. In winter months, consider heating the water troughs, or mix warm water into the cold water to avoid freezing. Goats are fanatical about their water, they will go thirsty before they drink water that is contaminated with debris or feces.

A Balanced Diet
I do not recommend mixing your own goat feed because most of are not trained as goat nutritionists. Commercial feed will contain a balance of all of the necessary nutrients. If you wish to feed organic, or as organic as possible, seek out an expert at a local feed or grain mill that is knowledgeable about the nutritional needs of a goat (one that does not treat goat nutrition as he/she would cattle, horses or other livestock).

General feeding guidelines:

Newborn and Young Kids – Begin life with colostrum, then milk, and introduce other feeds gradually (see Chapter 11, Raising Kids)

Does (Dry) – Good quality browse and pasture, supplemented with legume hay

Does (Pregnant or Lactating) – Good quality browse and pasture, supplemented with legume hay and grain (see Chapter 10, Breeding, and Chapter 12, Milking)

Bucks – Good quality browse and pasture, supplemented with legume hay

Wethers – Good quality browse and pasture, supplemented with a legume mix hay

Feeding Tips

To avoid Bloat and other illnesses, avoid sudden modifications to the diet. A gradual change allows time for the bacteria in the rumen to adjust. This rule also holds true when introducing goats to lush green pasture. Allow them to graze for short increments of time, preferably once a day for the first several days.

Goats do not tolerate diets low in digestible nutrients, even with that, they will pick through to obtain the optimal parts of the feed. It is not advisable to grind feed down. The goats will not be able to pick out the feeds that they need, which could lead to illness or death through overconsumption.

A grown goat will consume an average of 2-4 pounds of hay per day.

Legume hay (example: alfalfa) contains protein, vitamins and minerals that grass hay may lack. Legume hay is great for lactating does because it also contains calcium. When grass hay is the only option due to cost and availability, consider supplementing with feeds and minerals that replicate the nutritional balance of legume hay.

The quality of feed during the dry period (when does are not in milk or pregnant), is an important part of maintaining body condition and health year-round, which continues to affect the doe and the welfare of the kids throughout at least the following year. Keep in mind, maintaining the condition of a buck is just as important as providing a well-balanced diet to a doe. A buck works very hard during the breeding season; therefore, he should not become excessively fat, but must be in good condition all year round. Bucks have shorter expected life spans than does, which makes it even more important to ensure they have ultimate nutrition.

Other than hay, which the goats will munch on for long periods, the rule for feeding grain is to feed just enough to last an average of five minutes. Generally, a calculation of one cup of grain per goat, per feeding, is enough. I am not a fan of feeding grain free choice. Grain has a tendency to pick up moisture and spoil when exposed to damp climates, and the overeating of grain leads to rumen upset and grave illnesses. It is also more difficult to control weight gain, or to monitor how much any particular goat is eating when portions are not controlled.

Remember, in nature a goat survives on browse, leaves and vegetation from tall and deep-rooted plants, stems and weeds. These types of plants are not only nutritionally sound, full of life-giving minerals; they also provide roughage, which keeps the rumen and digestive system in top order. By providing roughage on your farm year round, through hay, you replicate this natural diet.

Supplementation

Provide loose minerals formulated specifically for goats. Minerals that are prepared for other types of livestock may not contain enough copper for a goat's needs.

Offer salt free choice, but when loose minerals are a part of the supplementation program, the mineral mixture may already contain salt. If this is a case, to keep the goat interested in the minerals, do not offer additional salt. Note: table salt designed for humans, or iodized salt, should never be a part of a goat's diet.

Offer sodium bicarbonate free choice. This helps to balance acids in the rumen, assists with digestive system functioning and stability, and decreases the chance of bloat.

Bucks, and especially wethers, are susceptible to Urinary Calculi (UC) (see page 117). Veterinarians and goat owners tend to go back and forth on how to prevent UC. Preventive measures often include offering sodium bicarbonate free choice, adding ammonium chloride to the diet, adding organic apple cider or unsweetened cranberry juice to the water. I personally do not feed grain to wethers, and I do not feed them alfalfa rich hay due to the calcium content. Instead, I feed a grass/alfalfa (or legume) hay mixture.

Molasses is a great source of energy. It contains trace minerals, and it adds to the palatability of feeds. Too much molasses in a diet can lead to rumen and digestive problems. For this reason, the percentage of molasses in feed should not be greater than 5%. An ill goat, or one weak from giving birth, will enjoy and get a boost from a small amount of molasses added to their water bucket (preferably warm water in very cold climates).

Feeders

Goats will crowd around and push their way to a feeder. A feeding station should be large enough to accommodate all of the goats within the paddock. It is important to monitor the goats while they are eating to ensure none are being crowded out and starved.

Goats, especially the youngest of the herd, tend to step into and walk through feeders. The less feed contamination, the healthier the goats. Keep this in mind when selecting feeders.

Feeders should be sturdy, well built, and immobile. To prevent mold, avoid placing hay racks in an area exposed to rain. Smaller hay racks, especially those that the goats reach up into, and those that they cannot step into, are preferred. Goats are natural browsers, it is much healthier for their digestive system and muscle development to reach up and pull hay downwards.

Entrapment is a concern when purchasing or building feeders. Goats can easily strangle or hang themselves in hanging feeders, chains, or wires.

Consider creep feeding, allowing young kids to feed separate from grown goats (see Chapter 11, Raising Kids).

If you live in a colder climate, and if you have electricity ran to your barn or feeding area, consider heated water buckets. This will prompt the goats to drink an adequate amount of water during the coldest of seasons, and will prevent you from having to break and dump ice out of the buckets. As a bonus, heated water troughs will keep you from the backbreaking chore of carrying warm water from the house to the barn!

Treats

Go light on feeding treats to your goats. I occasionally give animal crackers, soda crackers, peanuts in the shell, and even a treat of an apple or a carrot to a pregnant goat. Treats are fine as long as they do not substitute for daily nutrition, and if they do not prevent the goat from eating balanced feeds.

Very nutritious treats include leaves and branches from trees. Goats will eat bare leaves, and they will strip entire branches from live trees. They will quickly kill an apple, fruit, or evergreen type of tree. I suggest not enclosing these types of trees when setting up the goat paddocks. **Before feeding cuttings to your goats, ensure they are non-poisonous!** (See page 141, Poisonous Plants)

It is okay to feed non-poisonous leftover Christmas trees, but ensure they are organic, not sprayed with preservatives or chemicals.

Goats enjoy mouthing fabrics, plastic, and paper. I believe they also like noise – the crinkling, the rattling of plastic, paper, chains, gates, and fences. Goats are also very curious, which leads them to rip plastic and paper bags to shreds. The swallowing of fabrics and other items can cause rumen or intestinal blockage, leading to serious illness or death. When a goat eats an excess of any of these materials, watch them closely to ensure they do not show symptoms of illness, and if they seem to crave non-feed materials, ensure their diets are nutritionally sound. They may be lacking in minerals. Generally, though, they are just being goats!

Our Experience with Feeding (Advice from owners of other types of livestock.)

You can pick up valuable tips from experienced livestock farmers, even if they are not familiar with goat care, but they may not realize the dietary or housing needs of a goat.

A manager at a feed mill persistently attempted to persuade me to house and feed my goats as he did his cattle. I knew the feed was nutritionally wrong, but I still listened carefully to his advice. I filtered out what I could use as a goat farmer. In situations like this, it is smart to pay attention and learn what you can, but do not rely entirely on the information given. As with anything in life, use your common sense to pick out what pertains to you and your farm.

Our Experience with Feeding Browse (Nature's Minerals)

(Excerpt from AnniesGoatHill.com blog post)

Due to drought, the goats are struggling to find grass for grazing.

The goats are on hay, which is important to their well-being. A good alfalfa (or legume) blend is fed twice a day on our farm during times when pasture is scarce.

I also feed loose minerals, but a truly top-notch mineral for a goat is browse. Goats are browsers in their natural habitat (mountains and hillsides). A large shrub or tree, or deep-rooted plants, supply an immense scale of natural minerals to a goat.

I made a track around the farm this morning, slicing off branches from various trees such as apple, maple, and spruce.

The goats had a feast with the treats that were healthy as they could be.

Fall is a great time to gather fallen leaves for goats. Have you seen a goat chase after a leaf? Sometimes they catch them mid-air as they drift down from a tree. It really is an entertaining sight. Exercise, fun, and minerals combined into one great escapade!

When feeding browse, be aware that certain plants, especially ornamental yard plants, are highly poisonous to a goat. When feeding any cutting or plant material to a goat, always check to ensure it is not poisonous.

Our Experience with Feeding Poor Quality Hay

We once had a slim financial season. We were limited to a thinned out pasture and poor quality hay. The goats picked through the hay and ate what they thought was nutritious, leaving an enormous amount of waste behind. We soon realized that it would have been better to provide the hay that they needed, even if it had been in smaller quantities than they were accustomed to.

Chapter 8
Health and Wellness Care

Instinct and Observation

Goats are hardy animals; yet, their livelihood depends greatly upon our dedicated health and wellness care – with a focus on disease and parasite prevention.

As your goat herding experience grows in tenure, the more intuitive you become. You learn to trust your first instinct, which is the best indication that something is not right with an animal.

Plan to invest time with your goats every day, several times a day. Observation is one of the greatest teachers, therefore, it is time well spent. You will know the normal behavior patterns of your goats, which will help you to recognize anything out of the ordinary.

First Step in Treatment and Prevention

The first step in treatment and prevention begins with you. When you notice a goat is acting "off," which includes standing away from the herd, quieter than normal, walking with a different gait, lying down frequently when the rest of the herd is up or away, avoiding food, grinding teeth, and stretching frequently, assume that something is wrong, and check on the animal at regular intervals. An overly cautious goat farmer saves lives; one that ignores the first signs of trouble does not.

Separate the goat from the herd as soon as you begin treatment for a contagion. To avoid additional stress, keep the goat within sight and sound of the herd. Partitioning an ill or weak goat is always a smart decision, not only to prevent the spread of disease, but also because goats typically pick on a sick herd mate, often with so much force they wear the ill one down.

A Goat's Reaction to Illness and Pain

My conclusion is that goats put up with a great amount of initial pain and illness, which can mask external signals that anything is wrong to begin with. Much like a young child that resists slowing down when they are tired or ill, a goat's determination is to remain with the herd, eating and partaking in normal activities. They quickly give up once the pain and illness become too much to bear. Once they have gotten to this point, prompting them to get up and return to activity can be difficult. This is why it is important to catch an illness as early as possible, keep the goat on its feet, and continue working with the animal to ensure it regains strength and health.

Vital Statistics

- **Temperature** (anal) 102-104° Fahrenheit
- **Pulse** 70-80 beats per minutes (slightly more for kids)
- **Respiration** 15-25 per minute (slightly more for kids)
- **Rumen (stomach) movements** 1-1.5 per minute (will not be found in a very young kid)

Life Span
- **Does** – 11-12 years. Death is normally birthing related. Extend life span by retiring doe from breeding around the age of 8-10 years.
- **Bucks** – 8-10 years. Shorter than doe lifespan due to the enormous amount of energy they put into rut season.
- **Wethers** – 11-16 years. Early death can occur from Urinary Calculi (see page 117) for prevention.

Growth
Goats reach their full adult size when they are about three years old.

Teeth and Age
Goats do not have upper front teeth; they do have lower front teeth. They have both upper and lower teeth in the back of their mouth.

Estimate the age of a goat by their teeth:
- **Kids** – Goats are born with teeth that are small and sharp.
- **Yearling** – Goats lose their two middle front teeth around 12 months of age, which are replaced by permanent teeth.
- **Two Years Old** – Goats lose the two teeth next to the middle teeth around 24 months of age, which are replaced by permanent teeth.
- **Three Years Old** – At 3 years of age, goats have six permanent teeth, with only one pair of kid teeth remaining.
- **Four Years Old** – The set of eight front permanent teeth is complete.
- **Over Five Years Old** – It is more difficult to determine the age of a goat by their teeth after five years of age. Diet, pasture, and health care

are all factors in how quickly (if at all) teeth grind down, loosen, and eventually fall out.

The Medicine Chest

Illness will happen despite your best intentions. To avoid scrambling when the need pops up, this is a list of basic items to have on hand:

Supplements:
- Selenium (combined with vitamin E, injectable, prescription from veterinarian)
- Vitamin B complex (fortified with 100 mg thiamine per ml)
- Vitamin – iron – mineral feed supplement (purchase by the gallon, liquid found in equine section of store, helps to restore red blood cells in anemic animals)
- Vitamin C – 500 mg, up to 5 chewable tablets a day for adult goats when treating infections (supplement to boost immune system)

Antibiotics:
- Penicillin G
- Broad spectrum antibiotic

Pain Medication:
- Injectable pain medication (check with your veterinarian for availability)
- Anti-inflammatory (check with your veterinarian for availability)

Anti-Toxins:
- Tetanus anti-toxin
- CD&T antitoxin

Parasite Treatments:
- Coccidia treatment
- Dewormer
- Lice treatment

Intestinal/Stomach Treatments:
- Probiotic (gel or powder)
- Electrolyte
- Scour (diarrhea) treatment

Injuries:
- Blood stop powder
- Iodine (or disinfectant)
- Sprain wrap

Miscellaneous:
- Epinephrine (treatment for anaphylactic shock, see page 63)
- Hoof trimmers
- Thermometer
- Syringes (3 ML and 6 ML)
- Needles (20 gauge ½inch for kids, 20 gauge ¾ inch for adults)
- Storage container for used needles and syringes (Research your local regulations and disposal options. Never place used needles, also called sharps, in domestic or commercial trash or recycling bins. Keep out of the reach of children.)

Record Keeping
As your herd grows record keeping can be one of your most useful tool. For each goat, keep a log of illnesses, medications, treatments, vaccinations, worming, injuries, production/breeding, and the reasons for culling.

How to Administer Injections
There are two ways to administer injections:
- **Intramuscular (IM)** into the muscle
- **Subcutaneous (SQ)** under the skin

Goats metabolize injections given under the skin (SQ) better than they do intramuscular (IM). I believe this is because goats are generally not heavily muscled, so the injection given in a muscled area does not travel as far, nor does it always work its way through the goat's system. It loses effectiveness in the process. In most animals, not goats, IM injections work faster than SQ because they go directly into the plentiful muscle vs. the slower process of absorption into body tissues. Muscle damage can occur from IM injections, and there is a greater risk of hitting major blood vessels and arteries. For these reasons, I prefer SQ injections, even for most treatments labeled IM.

Injection sites are identified in a number of online illustrations (see Appendix, page 247). Do not give injections if you are untrained and inexperienced. Until you understand how to avoid arteries and major blood vessels (especially the back of the leg and neck areas), seek someone well versed to provide injections and instruction. I recommend visiting a veterinarian who will gladly allow you to observe the administering of rabies shots, or visit a farm for one-on-one instruction. I feel injections are an area of goat management taught and learned better in person, hands on, not from a book.

Subcutaneous (SQ) injections are administered under the skin, generally at the shoulder. Lift the loose skin and slide the needle under the skin, taking care not to hit the muscle or bone. Another area for SQ injections is the armpit behind the front legs. Lift the skin and inject down into the pocket under the skin. After giving the injection in this area, massage the injected site to reduce the possibility of a lump forming. Massage will also reduce any stinging that occurs from the treatment.

Always use good quality sharp needles. Do not share needles from goat to goat. Select the smallest diameter needle as possible. Larger diameter needles are necessary only for treatments that are very thick in consistency. Use a shorter needle for smaller and younger animals.

To prevent the needle from breaking or bending, tether or restrain the animal before treatment. To avoid sticking yourself accidently, consider wearing gloves, especially when injecting live serums or treatments.

A veterinarian, or a person well versed in injections, will also teach you how to fill the syringe, remove air pockets, and will give safety tips. If you are like me, I cannot tolerate the sight of human injections; however, I quickly learned not to be squeamish when treating my own animals. It is not a difficult task. It simply takes know how and confidence.

There is danger of anaphylactic shock, a sudden severe allergic reaction, any time injections are given. Epinephrine is adrenaline that immediately stops the deadly allergic reaction. It opens the airways and blood vessels. People with severe allergies to foods and insect stings carry epinephrine auto-injectors with them at all times. Consult with your veterinarian about the availability of this product. I have not experienced anaphylactic shock on my farm, but I know several farmers that have. When anaphylactic shock occurs, there is no time to fill a needle and syringe, so have it on hand!

Physical Therapy

Illness can lead to a weak goat, one that has difficulty sitting up, standing, or walking. Once a goat reaches this extreme stage they typically continue struggling to get up, which may wear them out further, or they may not try to get up at all.

If you can get a goat up on their feet, they have a much better chance of recovering from illness. This may require a makeshift sling, or a cart.

When a goat is unable to stand up, even with assistance, roll them from side to side frequently (no less than several times a day). This prevents the development of sores and stiff muscles. Frequent movement also relieves pressure on the rumen.

If you can get a goat on its feet, especially if you can entice them to eat fresh hay and drink water, I feel it adds to their strength, both mentally and physically. It is encouraging to the animal. You may need to be creative with a sling, or with a small stall you can stretch sheets or fabric across, which will hold up the middle section of a goat while they are doing therapy. Make sure it is safe, though, out of the reach of other goats, and set up in a manner that cannot cause injury or suffocation if the goat should slip.

To help strengthen muscles, consider a therapy sling for a young kid with weak legs. I use a sling made for use on a hanging scale. Attach the sling to a doorknob, or hang it from a sturdy hook in a safe environment. Position the kid in the sling so that their hooves touch the ground. Picture this, the kid is able to put weight on all four hooves and stand alone, yet they are able to relax into the sling when they become tired.

Common Diseases and Illnesses
The remainder of this chapter covers common diseases and illnesses, treatments, preventive measures, and personal notes from my experiences.

Clinical symptoms widely vary. A goat can have some, **but not all**, of the symptoms of an ailment, and some symptoms worsen as the affliction continues.

I have purposely withheld exact dosage amounts and schedules from this book. Drugs and medications vary by strength, and dosages are based on the extent of the illness, the condition of the animal, body weight, and determination of the exact cause. Use this chapter as a guideline. Always seek professional veterinary assistance.

***

Abortion

Abortion is the loss of offspring during pregnancy, or offspring that are born deformed or too weak to live.

Clinical Symptoms:
- Loss of offspring (born too early, deformed, or too weak to live)
- Discharge – blood and birthing materials

Treatment:
- See treatments for possible causes: Chlamydia, Listeriosis (Listeria), Vibrosis (Campylobacter), Toxoplasmosis, and to a lesser degree Leptospherosis and Salmonella
- Treatment may not be necessary if the cause is natural, such as injury or stress.

Prevention:
- Avoid rough handling and stress during pregnancy.
- Separate pregnant does from herd if fighting occurs.
- Check drug and medication labels for use during pregnancy, and administer careful

- To avoid transmitting pathogens to other goats, separate the doe from the herd until discharge stops. Handle and dispose of the fetus with care to ensure pathogens do not transmit to humans (especially pregnant women) or other goats.

Notes and Tips:

The second year we bred our does we experienced a rash of late pregnancy abortions. The vet advised us that it would "last for a season," and he said we would likely not see it occur in subsequent years. He was right. We determined that our abortions were due to Chlamydia. We treated accordingly and very seldom experienced abortions.

***

Acidosis (Grain Overload)

Acidosis results from consumption of large quantities of carbohydrates, most significantly grains. The overabundance of carbohydrates lowers the PH balance in the rumen, making it more acidic which causes it to function poorly.

Clinical Symptoms:

- Stomach discomfort
- Anorexia
- Teeth Grinding
- Muscle Twitching
- Cessation of normal rumen movements
- Diarrhea (off color, watery)
- Dehydration
- Increased pulse and respiration

- Bloat - distended tight left side (kicking or trying to chew at side)
- Lack of saliva
- Death

Treatment:
- Antacid preparation
- Electrolytes
- Antibiotic
- Water – plentiful, clean and fresh
- Pain medication
- Thiamine/Vitamin B complex fortified with thiamine

Prevention:
- Store grains where goats cannot reach them
- Avoid high grain diets, especially those high in corn content
- Introduce grains, or any new diet, over a period of 10-14 days (allows an microbial adjustment period)

Notes and Tips:
If you suspect that a goat has overindulged in grain, watch diligently for signs of acidosis and begin treatment immediately.

***

Anemia
An animal with a low red blood cell count is anemic. Anemia is a symptom, or an indication that an animal is in poor health. Various factors lead to anemia, but it most often occurs due to an internal parasite overload.

Clinical Symptoms:
- Pale inner eyelids
- Weight loss
- Rough dull coat
- Lowered milk yield
- Lethargy
- Bottle Jaw (see page 73)

Treatment:
- Treat the exact cause, the disease or illness
- Separate affected animal from the herd, and feed separate, until strength is gained
- Supplement with free-choice loose minerals
- Provide quality feeds balanced nutritionally for goats
- Supplement with non-poisonous plant materials – such as leaves and branches from trees
- Administer vitamin – iron – mineral supplements (see Supplements, page 60)

Prevention:
- Follow preventive measures for the exact cause, the disease or illness

Notes and Tips:
While the animal is recovering, monitor the eyelids to ensure they are returning to a bright or deep pink or salmon color (see FAMACHA, page 126).

It can take up to a year for an animal to return to a state of good condition. Improvement can happen quickly, but it takes time for the animal to gain weight, strength, and a healthy coat condition.

***

Anthrax

This extremely contagious disease (to animals and humans) is caused by Bacillus anthracis, a spore forming bacteria that remains alive in pastures for many years.

Clinical Symptoms:
- Sudden high temperature (104-108° Fahrenheit)
- Trembling
- Muscle tremors
- Depressed
- Anorexic
- Convulsions
- Staggering
- Difficulty breathing
- Bleeding from body openings
- Bloody diarrhea
- Sudden death

Treatment:
- Consult with your veterinarian for treatment (antibiotics and antiserum) and quarantine instructions

Prevention:
- Quarantine infected animals
- Vaccine (advisable in susceptible areas, consult with veterinarian)
- Pasture goats in dry areas (spores increase rapidly in damp conditions)
- Do not pasture goats in infected or suspected area

- Deep burial of infected carcasses (or burn if local regulations permit)
- Minimize fly and mosquito population

Notes and Tips:
Early treatment is the key to survival from this disease. Death may occur before the first symptom is noticed. The incubation period is 1-14 days.

As of this writing, Anthrax is rare in some areas, including most of North America, but it can infect anywhere. The disease is most common in agricultural regions of south and Central America, Europe, Asia, Africa, the Caribbean and the Middle East. There is a region in the United States, specifically in the state of Texas, where outbreaks have occurred.

It is a requirement to report any suspected cases of Anthrax to a veterinarian, who will then report to the CDC (Centers for Disease Control).

***

Blepharitis
Blepharitis is an inflammation of the eyelid resulting from allergies, parasites, or sun sensitivity.

Clinical Symptoms:
- Eyes red and swollen
- Eyelid spasms
- Loss of hair near eye (including eyelashes)

Treatment:
- Prescription from veterinarian
- Antibiotic ointment (purchased at farm stores) for mild cases

Prevention:
- Refrain from housing goats in dust laden areas
- Cleanliness of barn and bedding

***

Bloat (Frothy Mouth)

This condition occurs when gas or foamy substances build up in the rumen and cannot be expelled by burping or defecating.

Clinical Symptoms:
- Left side of abdomen distended
- Grinding of teeth
- Pawing the ground
- Kicking of abdomen
- Depression
- Difficulty breathing
- Foaming at mouth
- Death

Treatment:
- Encourage goat to walk – may help to expel gas
- Do not allow the goat to eat
- Antacids
- Store bought bloat treatment or drench with cooking oil (peanut, or other oil that the goat can taste and swallow is desired)

Notes and Tips:
My bloat experiences have involved young bottle fed kids. The culprit was milk replacer. Follow a strict bottle feeding schedule. If milk replacer is necessary, do not mix it too rich. Follow package instructions, it is better to make it too thin and gradually increase to the correct water to mix ratio. The thought process of feeding a rich milk replacer mix, or a very full bottle to help them gain weight, leads to rumen upset, illness and death.

There are two types of bloat – dry (gassy) and frothy (drooling at mouth). Dry bloat is eliminated by calming the goat and massaging their sides, urging them to walk to expel the gas, along with the administration of antacids. Frothy bloat requires the administration of substances (oils, manufactured bloat treatments) to break down the thick foam that forms in the rumen, which prevents the release of built up gases. Consult with your vet to determine the correct method of treatment.

The administration of bloat treatment, or oil, in very young kids may require tubing (see Tube Feeding, page 200).

Allow goats to graze on new or rich pasture only in short increments of time. This allows the rumen to adjust to the new diet. The feeding of hay before turning goats out to rich new pasture helps to decrease the chances of bloat.

***

Bottle Jaw (Big Head)

This occurs when the goat is heavily infested with worms or parasites.

A gastrointestinal parasite infestation causes a deficiency of proteins in the blood due to the thriving of parasites on the blood, and a decrease in nutrient supply due to loss of appetite in the animal. Proteins hold fluid in the blood. When blood proteins are lacking, the fluids leak and accumulate into areas such as the lower jaw, and lower abdomen, which causes swelling.

Clinical Symptoms:

- Anemia - mucous membranes are pale (especially the inner eyelids)
- Swelling under jaw or lower abdomen (when touched, feels as if jaw is or abdomen filled with fluids)
- Overall weakness
- Rough coat
- Scours (diarrhea)
- Lethargic
- Death

Treatment:

- Deworm
- Provide loose minerals, clean water, and nutritional feeds
- Vitamin B complex, fortified with thiamine
- Vitamin – iron – mineral supplement (see Supplements, page 60)

Prevention:

- Parasite prevention techniques (cleanliness, pasture rotation, see page 125 for additional preventive measures)

Notes and Tips:
Once the animal reaches the stage of edema in their lower jaw or stomach their life is in jeopardy. Treat swiftly to rid the goat of parasites, and to raise their blood levels (protein and iron), and to optimize rumen function.

Bottle jaw is particularly dangerous to young goats as they are in the process of building their immunity, growth and strength.

Some goats naturally look swollen under the chin. Get in the habit of checking the lower jaw if it looks puffy. This helps you to recognize the difference between fluid accumulation and a thicker jaw line.

The swelling from bottle jaw worsens in the evening, and sometimes does not appear in the morning at all. This is due to the goat holding its head in the grazing position, downwards, for most of the day, which causes the fluids to accumulate throughout the day.

***

Bronchitis
Bronchitis is inflammation or swelling of the bronchial tubes, the breathing passages between the nose and the lungs. The bronchial tubes may also become infected.

Clinical Symptoms:
- Deep and persistent coughing
- Cough may produce mucus
- Wheezing
- Fatigue

Treatment:
- Antibiotics

- Removing the cause – usually dust
- See treatment for Lung Worm and Mycoplasma, pages 131 and 104 (Bronchitis may be secondary to parasites and illnesses)

Prevention:
- Avoid housing goats in dusty areas
- Avoid overcrowding inside housing
- Dampen dusty feed (dampen prior to feeding, store feed dry to avoid mold)
- Avoid feeding dusty hay
- Avoid dusty bedding

Notes and Tips:
To guard the immune system, and to prevent further inflammation, irritation, or infection, treat respiratory ailments immediately.

***

Caprine Arthritis Encephalitis (CAE)
Caprine Arthritis Encephalitis (CAE) is a viral disease of goats caused by the Caprine Arthritis Encephalitis Virus (CAEV). CAE leads to failing health and physical ailments such as arthritis, or neurological symptoms that can lead to premature death.

Clinical Symptoms:
- Arthritis – most common in goats 6 months or older, mainly affecting the knee joints (unusual swelling)
- Encephalitis – most common in kids 2-6 months of age, with gradual paralysis
- Chronic pneumonia and cough
- Mastitis

- Weight loss
- Premature death

Treatment:
- Supportive care
- Cull

Prevention:
- Heat treat colostrum and pasteurize milk
- Purchase goats from CAE negative herds
- Maintain a closed herd

Notes and Tips:
As of this writing, testing for CAE is not accurate, both a false negative and a false positive can occur. The only accuracy level that an owner can reach is through continued testing with negative results, coupled with prevention.

CAE spreads through infected colostrum and milk, and to a lesser degree through saliva, urine, feces and blood.

CAE is more common amongst the dairy herds, but I have found no supportive data for this theory. More dairy kids are pulled from dam nursing, therefore, more are fed milk replacer, heat treated colostrum, and pasteurized milk.

There is no true treatment for CAE. Treatment lessens some of the initial symptoms, however, the disease is permanent and normally worsens as time goes on.

Some goats are carriers, but never show outward symptoms of CAE.

Arthritis and Encephalitis are common outcomes of a CAEV virus infection; however, other diseases also result, such as Pneumonia and Mastitis.

***

Caseous Lymphadenitis (CL, Cheesy Gland)

Caseous Lymphadenitis (CL) is a contagious and resistant bacterial infection spread from goat to goat through wounds and inanimate objects

CL abscesses appear in the lymph node areas, directly under the surface of the skin. Though much less common, abscesses may also form on the udder, lungs, and abdominal organs.

Clinical Symptoms:
- Abscesses in lymph node areas
- Higher morbidity rate later in life
- May lead to un-thriftiness, especially when abscesses affect internal organs

Treatment:
- Flush open abscess with iodine solution
- Formalin injections in abscess (contact veterinarian for instructions)

Prevention:
- Isolate animals with abscesses
- Vaccination (see **Notes and Tips**)
- Keep a closed herd
- Purchase animals from CL-free herds

Notes and Tips:
CL abscesses form in lymph node areas. They swell, lose hair, rupture and ooze out a thick pasty pus. The pus is a light greenish, yellow or cream color, with very little to no odor.

The veterinarian speaker at a goat health conference that I attended said, "If your herd is CL free, just wait, it will not always be CL free. Anyone that raises goats for a length of time will eventually experience it." His entire point was that CL is widespread, common, and testing can result in false positives. As of this writing, CL is incurable. Vaccinations are a preventive measure, but the disease can still pop into your herd.

The incubation period is as long as six months to a year, and CL bacteria can live in the soil for more than a decade.

CL research is ongoing; check with your veterinarian for testing, prevention, and the latest treatments.

CL is transmittable to humans when the pus from an abscess contacts broken skin. Always wear gloves when working with abscess materials. Burn the gloves, rags or paper towels that come in contact to prevent additional contamination.

***

Chlamydia (Chlamydiosis)
This infection is caused by the bacteria Chlamydia psittaci. It is a common reason for abortion in goats. There are many strains of this bacterium, some of which cause diseases such as Pneumonia, Pinkeye. and arthritis.

Symptoms:
- Abortions during the last 2-3 weeks of pregnancy
- Offspring born weak or early

- Abortions in young and first time pregnant does

Treatment:
- Antibiotics (oxytetracycline)
- Chlamydia vaccine

Prevention:
- Treat the entire herd (including non-pregnant does)
- Remove does from herd when they abort, until discharge stops
- Remove and bury aborted fetuses and birthing materials

Notes and Tips:
A buck infected with Chlamydia passes the bacterium from doe to doe.

Does infected with Chlamydia normally abort once, but not with subsequent pregnancies. If left untreated, goats continue to shed bacteria from their reproduction tract, which infects other herd members.

Use caution when treating a pregnant doe for this particular disease, or do not treat at all until she has kidded. Oxyetracycline may cause deformities in unborn kids.

Goats can be carriers of Chlamydia, especially bucks, with no outward signs of infection.

Use caution; wear gloves, when handling aborted fetuses and the birthing materials, both animals and humans can contract Chlamydia.

***

Coccidiosis

Coccidiosis is a disease caused by protozoan parasites that inhabit and damage the lining of the small intestine.

Clinical Symptoms:

- Diarrhea (with strong smell, may contain blood or mucous)
- Anemia
- Anorexia
- Weakness
- Dehydration
- Rough coat
- Poor growth
- Death

Treatment:

- Coccidiostats added to water or feed (especially those containing sulfa)
- Microscopic fecal exam
- Electrolytes (may require intravenous – IV – replacement of fluids if dehydration is severe)

Prevention:

- Feeds that contain coccidiostats
- Strict sanitation
- Minimize stress
- Avoid overcrowding
- Provide well ventilated dry housing, bedding, and feed areas
- Do not feed on ground
- Rotate pastures
- Provide clean water

Notes and Tips:
All grown goats carry a small number of coccidia occicites in their feces. As goats age they develop immunity, however, infection in adult animals can occur under extreme conditions.

Coccidiosis typically affects kids between the ages of 3 and 5 weeks old (the age when they begin eating more solid feeds and less milk).

The keeping of kids in crowded pens with adult goats increases the likelihood of infestation.

Coccidiosis occurs most often under two conditions: uncleanliness with poor manure management, and during warm and damp seasons. The parasite spreads orally through contamination of food, water and feeding surfaces. After ingestion of the eggs (or oocysts), they quickly multiply in the cells that line the gastrointestinal tract. Watery diarrhea is normally the first symptom, with an infective period of 14-17 days after symptoms begin. The life cycle of the oocysts continue within the goat during the infective stage, and the eggs shed when they reach sexual maturity. 2-3 days after the oocysts excrete into the environment, they reach a new level of maturity and infect a host if ingested. The entire life cycle for a coccidian parasite is approximately 21 days in length. Untreated coccidia results in oocysts that grow, reach levels of maturity, shed new eggs, and continue to destroy any remaining healthy intestinal cells.

***

Cystitis
Cystitis is an inflammation of the urinary bladder.

Clinical Symptoms:
- Frequent straining

- Small amount of urine
- Pus and blood in urine
- Depression
- Anorexia
- Excessive thirst
- Fever (when severity increases)

Treatment:
- Antibiotics (medication containing sulfa)
- Plentiful clean water

Prevention:
- Sanitation – overall cleanliness
- Management of overall health

Notes and Tips:
Cystitis and Urinary Calculi (see page 117) have similar symptoms. Urinary tract and bladder infections also produce inflammation and infection, with similar symptoms.

Consult with your veterinarian for diagnosis and treatment.

***

Diarrhea (Scours)
Diarrhea is a symptom, an indication that something is not right with an animal.

Clinical Symptoms:
- Runny, loose, or pasty feces

Treatment:
- Treat the cause, never treat the diarrhea as a single ailment (unless you are sure it is caused by a slight and correctable change in feed or environment)
- Electrolytes (young kids especially – to prevent dehydration)
- Probiotic

Prevention:
- See prevention of the specific disease or ailment (the cause)

Notes and Tips:
There are treatments on the market that effectively control scours. These products contain antibiotics, marketed for pigs, and are helpful for the control of goat scours. I cannot stress enough, though, determine the reason for the diarrhea, and treat that particular ailment first. Keep in mind, when the body is ridding itself of toxins, a small amount of diarrhea can be good. Diarrhea becomes dangerous only when it is profuse, ongoing, and leads to dehydration.

***

E.coli (Collibacillosis)
E.coli is a disease caused by the common bacteria, Escherichia coli.

Clinical Symptoms:
- Salivation increase (outside of mouth may appear wet or bubbly)
- Diarrhea (thin, light in color)
- Loss of appetite

- Dehydration
- Hunching up (due to stomach pain)
- Fever (may or may not have fever)
- Cold mouth (in later stages)
- Bloat (slightly bloated appearance)
- Death

Treatment:
- Antibiotics
- Electrolytes
- Fluids

Prevention:
- Sanitation (clean and dry housing and bedding, sanitary bottles and nipples)
- Avoid overcrowding of animals
- Avoid cold milk and irregular bottle feeding (see Bottle Feeding, page 190)
- Avoid stress
- Ensure newborns receive colostrum at birth
- Dip naval of newborn goat (see page 173)

Notes and Tips:
E.coli is most common in kids 1-4 days of age, occurring less frequently in kids up to a month old.

With a short-lived incubation period, 1-2 days in young goats, toxins from the bacteria form in the internal organs, and death quickly follows, usually within 12-24 hours of the onset of diarrhea.

Wash your hands thoroughly after handling a goat infected with E.coli. Bacteria can be transmitted to humans, especially affecting young children and the elderly

Due to similar symptoms, E.coli is often confused with Floppy Kid Syndrome (see page 87).

***

Enterotoxemia (Overeating Disease)

Enterotoxaemia is caused by two strains of Clostridium perfringens bacteria - types C and D.

Clinical Symptoms:

- Convulsions
- Anorexia
- Abdominal pain
- Lethargic
- Profuse watery diarrhea (often mixed with blood or mucus)
- Fever 105° Fahrenheit
- Uncoordinated movements
- Sudden death

Treatment

- C&D antitoxin
- Penicillin
- Fluids
- Pain medications

Prevention

- Avoid high carbohydrate diets (too much grain with little forage)
- Avoid sudden change in feed (make gradual changes)
- Avoid sudden exposure to fresh green pasture (feed hay beforehand and limit time)
- Avoid overcrowding

- Avoid the overfeeding of milk and replacer to kids
- Vaccinate Clostridium perfringens Type C and D, and Tetanus (CD&T).

Vaccination Schedule (adults):
- Initial vaccination (if more than one year has passed between vaccinations), followed by booster in 21-28 days
- Booster (annual - if less than one year between vaccinations, may booster up to 2 times a year)
- Booster (pregnant does, 20 to 30 days prior to expected delivery date – provides antibodies in colostrum)
- Booster every 3-6 months if infections of Enterotoxemia have occurred in the herd

Vaccination Schedule (kids):
- Initial vaccination 10-12 weeks of age (if dam received booster 20 to 30 days prior to birth)
- Initial vaccination at birth (if dam did not receive booster 20 to 30 days prior to birth)
- Booster (within 21-28 days) followed by annual booster

Treatment:
- C&D antitoxin
- Antibiotics (penicillin)
- Fluids
- Electrolytes
- Pain medications (contact veterinarian)

Notes and Tips:

As a goat matures they are less susceptible to Enterotoxaemia, however, they always carry a low percentage of Clostridium perfringens bacteria in their small and large intestines. When a sudden change in feed occurs, or when a goat overloads on carbohydrates, higher levels of starch, sugar, and protein reach the intestines, which causes the Clostridium perfringens to increase. This leads to the release of toxins, which causes this deadly disease.

The vaccine for Enterotoxemia is normally a three-way type, which includes protection for Clostridium perfringens (types C and D), and Tetanus (T).

***

Floppy Kid Syndrome (FKS)

Floppy Kid Syndrome is not caused by a specific organism or bacteria. It occurs when the digestive system stops working correctly when the kid overindulges on milk.

Clinical Symptoms:

- Affects kids between 2-10 days old
- Temperature normal
- Depression
- Weakness (flaccid paralysis as the disease progresses)
- No diarrhea
- Abdominal distension (as the disease progresses)
- Inability to use tongue (kid can swallow, but appears unable to nurse)

- Does not respond to antibiotics or selenium supplementation
- Symptoms come in waves (appearing to be better, then worse, and repeat)
- Death

Treatment:
- Remove milk for 24-36 hours
- Pain medication (may be needed in advanced cases, consult with veterinarian)
- Baking soda (sodium bicarbonate, to neutralize stomach condition, see **Notes and Tips** below)

Prevention:
- Do not overfeed milk

Notes and Tips:
The scenario goes as such: the kid is eating well, appears healthy, and has been happily bouncing all over the place. The next thing you know, the kid is weak, or to the point that they are flaccid, unable to stand or sit up. This situation is much different from a newborn kid that is born weak and unable to nurse or stand from the first moments of life. Floppy Kid Syndrome (FKS) has one cause – the ingestion of too much milk.

Resist the urge to force feed milk to a weak kid, one that does not want to nurse, or one that seems to be unable to suckle. Milk will only make FKS worse.

The first step in treating FKS is to remove the kid from milk or replacer, substitute with a baking soda (sodium bicarbonate) and water mixture: 1teaspoon of baking soda dissolved and mixed completely in 8 ounces of water. If the kid will not nurse on this mixture, you may need to syringe feed. Draw the liquid into the syringe and slowly administer it into the kid's mouth. Caution – do not force-feed. Liquids can enter the lungs and bring on infection (Pneumonia). If the kid readily swallows the liquids, continue syringe feeding, slowly, until the belly rounds out (as if he is full). If the kid is very weak, unable to hold its head up, Tube Feeding will be necessary (see page 200). Consult with your veterinarian for further instructions.

Administer the baking soda mixture 2-3 times within the first 3-6 hours.

After the treatments, the first feces will appear solid and normal; it will then turn to diarrhea. Do not treat the diarrhea. Remember, the kid ingested too much milk and the body is ridding itself of the fluids that blocked the intestines.

After administering treatment, and after the feces pass, begin replacing fluids with water or watered down electrolytes.

After the kid appears to have recovered from FKS, no sooner than 24-36 hours after recovery, begin feeding milk. If you are bottle feeding, dilute the milk (1/4 milk to ¾ water) and gradually increase the milk/water ratio over a period of several days. Again, do not overfeed milk (see Bottle Feeding schedule, page 190).

If the kid is not on a bottle, you may return him to his dam, but I recommend doing so only in small increments of time until his body adjusts once again to the richness of whole milk. To ensure the kid does not overindulge, especially if it is a single kid on the dam, or if she is a proliferous milker, consider milking her several times a day. This will leave enough milk for the kid, but not so much that it will make him ill.

One final tip on FKS - before the kid is treated he may appear to improve, followed by bouts of him going flaccid, slipping into a coma within 15-20 minutes, and once again returning to a state where he appears to have recovered completely. These symptoms commonly come in waves, and the only cure comes from swift treatment. When no treatment is given, death occurs within 24 hours.

***

Foot Rot and Foot Scald
Foot rot and foot scald are similar diseases caused by microorganisms – Dichelobacter nodosus and Fusobacterium necrophorum. Depending upon the strain, various areas of the foot and hoof are affected.

Clinical Symptoms:
- **Foot Rot** – Outer shell of hoof separates from inner sole, erosion of tissue between sole of toe and hard outer hoof (contagious, severe cases: pus and foul odor, loss of appetite, hoof deformity)
- **Foot Scald** – Skin between toes is pink to white in color, raw, moist, sensitive to touch (not contagious, precludes foot rot)

Foot Rot and Foot Scald:
- Foul odor (occurs with both diseases, but not as often in less severe Foot Scald cases)
- Lameness and reluctance to walk
- Limping

- Holding hooves off of ground
- Grazing on knees
- Reduced weight due to inability to graze, or, weight gain due to lack of exercise

Treatment:
- Isolate to ensure goat is able to walk, eat, and drink fluids without competition
- Antibiotics
- Footbath with zinc or copper sulfate
- Treat hooves with copper sulfate, zinc sulfate, or a 7% iodine solution
- After treatment, place goats in dry environment for at least 24 hours
- Trim hooves to expose infected tissue to oxygen

Prevention:
- Avoid overgrown hooves (trim when needed)
- Avoid prolonged exposure to warm moist areas such as wet pasture, pens, or muddy soil
- Quarantine new animals (with footbath upon arrival and for two weeks)
- Avoid contamination - can be carried on boots, tires, equipment, feeders or hands
- Administer a foot bath to each goat after returning from a show, or any time they have visited areas where goats from other farms are present
- Maintain clean and dry pens and barn

Notes and Tips:
If either disease, Foot Rot or Foot Scald, is not treated, the hooves may become permanently infected.

The microorganisms that cause either disease can live in the soil for several weeks once it is contaminated.

Some animals are pre-disposed to these diseases through genetics. In these cases, selective breeding and culling is a preventive measure.

When dealing with lameness, look beyond Foot Rot and Scald to rule out other possible causes.

As of this writing, a vaccine for either disease is not available for goats. Consult with your veterinarian for vaccine and treatment options.

***

Hypocalcemia (Milk Fever)

Hypocalcemia is a calcium deficiency most often seen in a pregnant or nursing doe. The deficiency results when the doe is no longer able to metabolize and absorb calcium from her diet.

Clinical Symptoms:
- Lethargic
- Eats less or stops eating
- Bloat – mild
- Constipated
- Wobbly gait – especially in rear legs
- Staggers while walking
- Difficulty standing from a sitting position
- Muscle trembling
- Decreased body temperature
- Shivering after milking
- Weak uterine contractions
- Confusion
- Stops ruminating, urinating, or defecating (later symptoms)

- Death

Treatment:
- Calcium gluconate
- CMPK or Cal MPK (over the counter solution containing calcium, potassium, magnesium, sodium, borogluconate)
- Vitamin B complex, fortified with thiamine (to increase appetite)
- Milk less frequently (to slow down milk production)
- Molasses in warm water (encourages doe to stand and provides energy)
- Swift treatment helps to avoid death.

Prevention:
- Do not over feed calcium in final 30 days of pregnancy. If feeding legume hay, such as alfalfa, do not feed high calcium grain or supplements that contain calcium.
- Exercise
- Proper Nutrition
- Provide free choice minerals
- Maintain correct weight (not too thin or fat)

Notes and Tips:
The exact cause of Hypocalcemia is difficult to pinpoint. A doe experiences sudden increases in calcium and phosphorus requirements when she is pregnant or lactating. Even though she is eating feeds high in calcium, and she may have enough calcium in her bones, the disease can set in. It is thought that an excess of calcium fed during the last trimester of pregnancy can lead to the imbalance.

It is imperative to watch a doe late in pregnancy, or while in milk, for signs of weakness. If it appears that she cannot stand, or if she staggers when she walks, or shivers after she is milked, treat this as suspected Hypocalcaemia as early as possible. If left untreated, it can be lead to Enterotoxemia, Mastitis, retained placenta, and death.

Hypocalcemia is often confused with Ketosis (Pregnancy Toxemia). Ketosis occurs during pregnancy. Hypocalcemia can occur both prior to and after giving birth. The differences between the two conditions are, Ketosis results in the doe metabolizing her body fats in an effort to maintain nutrient and energy levels. Hypocalcemia results in the doe not metabolizing calcium from feeds or from internally stored calcium.

***

Johne's (Paratuberculosis)

Johne's (pronounced Yo-nees) is a gastrointestinal disease caused by the bacterium Mycobacterium avium sub species: paratuberculosis (MAP).

MAP invades the intestines, causing the walls to thicken, which prevents nutrients from absorbing. Essentially, the goat can eat healthy amounts of feed but still starve to death. In later stages of the disease, MAP invades the muscles and other organs.

Clinical Symptoms:
- Rapid weight loss
- Diarrhea
- Unthrifty appearance
- Rough coat

Treatment:
- No known cure

Prevention:
- Test your herd
- Keep a closed herd
- Separate or cull animals that test positive
- Purchase Johne's tested goats
- Do not borrow other people's goats, or use their pastures
- Graze in non-infected pastures
- Purchase milk and colostrum from tested Johne's free herds, and heat treat colostrum, pasteurize milk
- Cleanliness of kidding areas
- Avoid feeding on ground
- Cleanliness of water troughs and feeders

Notes and Tips:
As of this writing, there are no known cures or vaccines for Johne's.

Johne's normally infects in the first few months of life. The disease seldom infects a grown goat unless their immune system has been compromised.

An infected animal can appear healthy for years (into adulthood). Then they begin to lose weight and later develop diarrhea.

Due to the unthrifty appearance and weight loss of the infected animal, Johne's is often misdiagnosed as an internal parasite infestation.

Infection normally spreads through pasture, water, and milk contaminated with feces containing MAP bacteria. The unborn can also be infected through the placenta.

***

Ketosis (Pregnancy Toxemia)

Ketosis is a metabolic disease that occurs one to three weeks prior to kidding. When the pregnant doe cannot ingest enough nutrients to meet the glucose requirements of both the kids and herself she metabolizes her own body fat. Ketones are produced which enter the blood stream and are dangerous at high levels.

Clinical Symptoms:
- Eats less or stops eating
- Depression
- Separation from herd
- Slow rising or too weak to stand
- Dull eyes (can lead to temporary blindness)
- Muscle tremors/seizures
- Staggering
- Head pressing
- Teeth grinding
- Rapid breathing
- Breath smells sweet (caused by ketones)

Treatment:
- Treat early (before the doe is too weak to stand)
- Propylene Glycol
- High energy electrolytes
- Molasses in warm water
- Veterinary treatment (IV, induced labor, C-section)

Prevention:
- Feed quality hay and forage
- Feed correct amount and type of grain

- Keep stress to a minimum
- Avoid overweight (caused by overfeeding)
- Exercise
- Separate doe from herd if she is struggles with eating
- Ovoid overcrowding

Notes and Tips:
The key to treating this disease is catch it early, treat as soon as the symptoms are realized, continue diligent treatment, and involve a veterinarian as soon as it is recognized that treatment is not be working.

Note: Rapid fetal growth occurs in the final two months of pregnancy, which leaves limited rumen space. The doe may eat less because she feels full, especially if she is carrying large kids, or more than two. Separate the doe from the herd if you notice that she is eating just a small amount of feed and then walking away from the trough. Separation allows her to eat the ration that she needs at her own slower speed, and it helps you to monitor the amount that she does eat.

Proper weight management prior to breeding and during gestation is important. The doe should not be excessively thin or overweight prior to breeding, nor should she gain excessive weight during pregnancy, and she should not lose weight during the final trimester of pregnancy.

Ketosis (Pregnancy Toxemia) is often confused with Hypocalcemia (see Milk Fever, page 92).

***

Lice

Goats can host two types of lice, biting or sucking. Both cause intense itching. Symptoms, such as anemia, are relative to the exact species of lice and the degree of infestation.

Clinical Symptoms:

- Scratching, rubbing and itching
- Biting at self
- Raw areas
- Loss of hair
- Scabs
- Anemia

Treatment:

- Powder or applicable over the counter treatment
- Dewormer that contains lice inhibitor
- Remove nits by brushing
- Treat on day 1 and repeat in 10 to 14 days to kill newly hatched lice

Prevention:

- Sunshine and outdoor exercise helps (does not totally prevent)
- Dewormer that contains lice inhibitor

Notes and Tips:

People are not susceptible to goat lice.

Goats are prone to lice during long winters when they spend more time in the barn with straw as bedding. As soon as warmer temperatures and sunshine arrive, the fight to control lice eases off naturally.

Lice and their grayish nits (eggs) can be seen deep in the goat's coat by using a magnifying glass, especially along the backbone.

The determination of the exact species of lice requires a microscope. Sucking lice have large heads, compared to the smaller heads of biting lice. Unless the infestation is severe, with large areas of hair loss, bleeding scabs, or very raw areas, both types of lice are controlled by following the simple treatment steps listed above. Consult with a veterinarian in severe cases.

***

Listeriosis (Circling Disease)
Listeriosis is life threatening brain-stem disease caused by the bacterium Listeria monocytogenes.

Clinical Symptoms:
- Depression
- Anorexia
- Food packed in cheek
- Decreased milk production
- Leaning, stumbling, or circling in same direction (in severe cases – falls down on side with involuntary running movements)
- Seizures
- Facial paralysis (may have food packed in cheek)
- Ear droop
- Slack jaw, salivation or drooling
- Swelling
- Fever
- Diarrhea, abortion and pregnancy toxemia
- Death

Treatment:
- Antibiotic
- Anti-inflammatory
- Dexamethasone (veterinarian will prescribe to reduce brain swelling)
- Electrolytes and plentiful water
- Vitamin B Complex (fortified with Thiamine)

Prevention:
- Avoid moldy, spoiled hay and feed (do not feed silage)
- Avoid sudden changes in feed
- Cleanliness - housing, feed, pastures and water
- Treat open wounds to prevent bacteria from entering
- In cases of abortion – **discard fetus and birthing materials, handle with gloved hands (humans can contract Listeriosis)**

Notes and Tips:
Treatment must be rapid to avoid death!

Remove kids from infected dam, heat treat colostrum, pasteurize milk or feed milk replacer.

Two forms of Listeriosis affect goats, the first of which is encephalitic – with the clinical symptoms mentioned above. Diarrhea is present only in the form that causes abortion and can lead to pregnancy toxemia.

The symptoms of Listeriosis closely mimic those of Goat Polio (see Polioencephalomalacia, page 110).

*****

Mange

Mange is an intense itchy dermal condition caused by the mange mite. The three types of mange mites that affect goats are scarcoptic, psoroptic, and chorioptic.

Clinical Symptoms:
- Itching
- Crusts or scabs on skin that have wrinkled, thickened or are hairless
- Flaky dandruff
- Skin infection
- Poor semen production in bucks (advanced cases)

Treatment:
- Vet must test sample to determine treatment based on (one of three) types of mange mites
- Sulphur or lime sulphur dip
- Dewormer

Prevention:
- Treat all animals involved, including livestock guardian dogs
- Quarantine new animals

Note and Tips:
Treatment may take up to 3 weeks to rid animals of mange.

A veterinarian must make an accurate diagnosis of the type of mange mite affecting the herd microscopically.

The scarcoptic mange mite is the most difficult to eradicate because it burrows into the skin, making tunnels in which it lives and lays eggs.

Mange may disappear in the spring, appearing as if treatment has been completely successful, only to return once again in the winter when the temperatures drop and exposure to sunshine lessens.

***

Mastitis

Mastitis is inflammation and congestion of the mammary gland, the udder, resulting from a bacterial infection, toxin or physical trauma.

Clinical Symptoms:

- Warm, swollen and painful udder
- Milk abnormal in consistency and/or color
- Blood in milk
- Not eating
- Fever
- Lethargic
- Udder changes color

Treatment:

- Antibiotics (injections and formulas inserted into teats)
- Antibiotics (traditional injections for advanced cases)

Prevention:

- Improve sanitation (housing, pasture, cleaning of teats and udder prior to milking, and milking equipment)
- Determine protein levels in the diet, too much protein lowers copper absorption which can lower immunity to infections
- Proper adjustment of milking machine

- Determine calcium/magnesium ratio in diet, imbalance places the udder at risk of bacterial invasion
- Dip the teats to close orifices after milking
- Avoid rough handling of the udder and teats
- Do not strip the udder of milk. Milk only until the stream comes to a near stop.

Notes and Tips:

Blood or clumpy milk, or a congested udder, can occur with or without the presence of Mastitis (see Milking, Chapter 13).

Take a sample of the milk for testing as soon as Mastitis is suspected. Most farm supply stores carry Mastitis test kits, preferably the California Mastitis Test kit.

If the infection has gone septic, i.e., has gone into the bloodstream, permanent damage, extreme illness, and death of the doe can follow. Be aware of these symptoms: fever, anorexia, depression, and lethargy.

Suspect an infection of Mastitis when offspring fail to thrive, and/or the dam does not want the kids to nurse. Never allow kids to nurse on an infected udder. After removing the kids from the dam, watch for signs of illness and treat them with antibiotics as a precautionary measure.

Most cases of Mastitis consist of similar treatments and symptoms; however, the type called black Mastitis (or blue bag) attacks very suddenly and severely. Within an hour, or two, the infection causes the udder to change color from red to purple to black. Unless immediate treatment is administered, the udder will be damaged, and may need to be amputated. Death from this type of Mastitis is swift. If you suspect black Mastitis, contact your veterinarian immediately for instructions and treatment.

***

Metritis
Metritis is an inflammation of the uterus.

Clinical Symptoms:
- Slight signs of lowered health, such as, not eating or drinking well, walking slowly with head down, standing apart from herd
- Straining
- Swollen vulva
- Fever (in advanced cases)

Treatment:
- Vitamin A & D
- Antibiotic (Penicillin)

Prevention:
- Treat doe with antibiotics after assisted kidding
- Infection may be spread from buck, watch for signs of illness

Notes and Tips:
Milk from a doe that has Metritis may be contaminated. Heat treat the milk before feeding it to kids, and pasteurize or avoid the milk for human consumption.

Metritis can occur at any time, not just after kidding.

***

Mycoplasma

Mycoplasmas are slow-growing virus-like infections. These microbes have no cell walls, which allows invasion of tissues and organs. There are literally hundreds of different Mycoplasma subtypes and strains.

Clinical Symptoms:

- See the list of **five basic problems** (respiratory, arthritis, mastitis, conjunctivitis and septicemia) below.

Treatment:

- Antibiotics
- Tetracycline
- Bottle feed kids if nursing from dam with suspected or confirmed Mycoplasma infection
- Prompt and swift treatment
- Isolate suspected cases immediately
- Culling of all animals infected with mycoplasma infection (some authorities have suggested this)

Prevention:

- Prompt and swift treatment
- Isolation of infected animals
- Culling

Notes and Tips:

At the time of this writing, a test that detects Mycoplasma has been developed, but it is not readily available.

I strongly recommend seeking veterinarian assistance immediately for diagnosis, treatment, and prevention. Some treatments lead to the growth of additional infections.

In goats, mycoplasma leads to **five basic problems:**

Respiratory

- Affects adult and young goats
- Extremely contagious
- Results in high percentage of death
- High fever
- Head lowered or extended
- Arched back
- Increased lung sounds and respiratory rates

Arthritis

- Affects young goats (generally 2-4 weeks old)
- Limping
- Hunched back
- Laying or squatting down and crying out in pain
- Swollen and hot joints
- Blindness (in some cases)
- Death within a few hours

Mastitis

- Signs of Mastitis in dam (see page 102)
- Nursing kids may show signs of pneumonia, arthritis, conjunctivitis, septicemia, fever, weakness and extremely high death rate

Conjunctivitis

- Yellow discharge
- Difficult to differentiate between other eye infections unless it occurs with additional Mycoplasma symptoms

Septicemia
- The infection penetrates through the entire body and will resemble Arthritis, and may result in any of the symptoms listed above.

***

Pinkeye (Keratoconjunctivitis)
Infectious chronic Pinkeye is a contagious disease often resulting from a Chlamydia infection (see page 78), or Mycoplasma (see page 105). Non-infectious pinkeye is often caused by over-exposure to very bright sunshine and billowing dust.

Clinical Symptoms:
- Red, swollen, running eyes
- Opaque cornea (dark park of eye)

Treatment:
- Isolate goat from herd
- Triple antibiotic ophthalmic ointment
- Oxytetracycline (in severe cases)

Prevention:
- Avoid stress
- Sanitation of barn and paddocks, reduce dust
- Avoid overcrowding
- Fly prevention

Notes and Tips:

Pinkeye occurs most often in warm or hot weather. It is spread by flies, and through body contact, and it can spread from farm to farm.

Some goats recover from pink eye without treatment, before it is known that the disease has set in, however, severe untreated cases can cause permanent blindness.

Infected animals are more sensitive to bright sunlight. Isolate the affected animal in a shaded area.

In severe cases, if the eye appears to be ulcerated, the affected eye may need to be patched to avoid additional irritation. Ulcerated eyes can become permanently damaged when treated incorrectly. Contact your veterinarian for diagnosis and treatment.

Pinkeye is specific to species. Pinkeye in cattle is not the same infection as pinkeye in goats.

Pinkeye is highly contagious to both animals and humans. Wear gloves when treating this disease, and begin treatment as soon as possible to prevent spreading within your herd.

***

Pneumonia

Several organisms cause pneumonia in goats, the most common are Pasteurella multocida and Pasteurella haemolytica.

Clinical Symptoms:
- Fever
- Nasal and eye discharge (nasal discharge is not clear)
- Anorexia
- Difficulty breathing

- Persistent cough (pain, hunching up later in illness)
- Drowsiness

Treatment:
- Antibiotics
- Oxytetracycline (non-pregnant does)
- Anti-inflammatory (to reduce fever, prescription from veterinarian)
- Electrolytes and plentiful clean fresh water
- Veterinarian culture

Prevention:
- Quarantine (can be contagious)
- Vaccine (contact veterinarian for strain-specific availability)
- Avoid stress
- Avoid confinement in high ammonia or dusty conditions – keep the barn and bedding clean (minimize any environmental condition that can irritate the lungs)
- Adequate ventilation in housing – avoid continual and direct cold and damp drafts
- Good nutrition and plentiful clean water
- Avoid overcrowding

Notes and Tips:
Pneumonia and related respiratory problems are more common in kids, but the infection affects goats of all ages.

It is not just a cold weather ailment; summer, with hot days, high humidity, and cool nights may result in Pneumonia. Pneumonia strikes year round.

Pneumonia caused by Pasteurella haemolytica can kill a goat in a very short time period, in less than 12 hours, and normally in 5 or 6. Unfortunately, there are few symptoms until the final few hours of life. The first symptom is commonly a very high fever. For this reason, consult with your vet for culture and treatment.

Pneumonia caused by Pasturella multocida will have clinical symptoms that may or may not include fever, which worsen when left untreated, but death is not as sudden in this strain. Contact your veterinarian, to perform the necessary tests and provide treatment.

***

Polioencephalomalacia (Goat Polio)
The cause of Polioencephalomalacia is thiamine deficiency, which affects the brain cells.

Clinical Symptoms:
- Incoordination
- Weakness
- Tremors
- Depression
- Blindness
- No fever
- Diarrhea
- Muscle contractions and seizures
- Comatose and death within 24-72 hours

Treatment:
- High doses of thiamine (Vitamin B1) injections
- Anti-inflammatory (in severe cases)
- Isolate to provide access to feed and water

Prevention:
- Avoid moldy hay or feed
- If water or feed is high is sulfur, supplement feed with Thiamine (Vitamin B1)
- Avoid overfeeding of grain
- Avoid fish products and corn distillation (ethanol) by-products that contain high levels of sulfur.

Notes and Tips:
Goat polio most often affects kids in the 2 – 8 month old range.

Injectable Fortified Vitamin B complex (must state fortified on the bottle) is a key supplement to keep on a goat farm. I recommend the complex that contains 100 mg of thiamine (Vitamin B1) per ml. A goat cannot overdose on thiamine; the excess will shed off in urine.

Sulfur inhibits the absorption of thiamine. Water and feeds containing high amounts of sulfur can lead to thiamine deficiency, which causes Polioencephalomalacia. In severe cases, consider testing the water supply. In less severe cases, provide feeds supplemented with thiamine (vitamin B1).

Symptoms of Polioencephalomalacia closely mimic those of Listeriosis (see page 99).

***

Ringworm
Ringworm is a fungal disease, not a worm at all.

Clinical Symptoms:
- Rounded patch of hair, surrounded completely by a hairless ring
- Irritation and rubbing

Treatment:
- Wash the area with a topical skin disinfectant
- Apply an anti-fungal cream or spray
- Lime sulphur concentrate dip (contact veterinarian for treatment and instructions)

Prevention:
- Quarantine affected goat
- Cleanliness and removal of damp materials from barn, bedding, and paddocks

Notes and Tips:
Ringworm is one of the most common skin problems in goats. It is **highly contagious to humans** and animals. Prevention includes quarantine of the animal, and the **wearing of gloves while handling and treating**.

Ringworm will run its course. It pops into a herd mainly during long wet seasons, when animals tend to spend time inside the barn. The fungus thrives in soil, and other dark and damp materials – such as manure and bedding.

***

Sore Mouth (Orf, Contagious Ecthyma)
Sore mouth is a highly contagious skin disease caused by a virus in the pox family. It affects animals of all ages, but primarily affects the lips and noses of young animals.

Clinical Symptoms:
- Red spots on the lips or nose which form blisters that burst and form ulcers (can also appear on ears, feet, scrotum, teats and vulva)

Treatment:

- Healing can occur without treatment in mild cases and the scabs fall off within three to four weeks (see **Notes and Tips** regarding lesions on the teats and udder)
- Secondary infections may require broad spectrum antibiotics
- Severe cases may require application of antibiotic cream to affected areas of skin

Prevention:

- Vaccine (see **Notes and Tips**)
- Quarantine affected animal from herd
- Quarantine new animals from herd for a minimum of 30 days
- Clean and disinfect contaminated pens, discard contaminated bedding

Notes and Tips:

A vaccine for Sore Mouth is available. It is a live vaccine, which can cause the animals in your herd to develop sore mouth blisters and ulcers if they have not previously been exposed to the virus. Discuss vaccination availability, and whether it is a necessary preventive measure, with your veterinarian.

The teats and udder of female goats can be affected by Sore Mouth if infected kids are nursing. It can lead to Mastitis if it goes untreated. If milking equipment is being used, milk the infected goat last and thoroughly disinfect all milking equipment after use. Use an antiseptic udder salve on the teats and udder to control bacteria until healing occurs.

Young kids may quit nursing and eating due to the lesions. If this should occur, treat the lesions with antibiotic creams or sprays, and ensure the kid receives adequate nutrition.

As of this writing, there is debate amongst goat farmers if Sore Mouth, or the vaccine, leads to abortion. The initial reports of abortions occurring were from large herds that were rounded up, which stressed the animals and could have added to a lack of immunity to disease, both of which increase the likelihood of abortion.

Animals are immune to the disease for one year following a Sore Mouth infection. It has not been proven, but some animals seem to be immune for life. Keep in mind, the virus has been found live in dry scabs up to twelve years after they had been shed.

Use caution, **wear gloves at all times, sore mouth may be contagious to humans**.

***

Tetanus (Lockjaw)
Tetanus is a bacterial disease caused by the bacterium Clostridium tetani.

Clinical Symptoms:
- Rigid muscles (especially the mouth and jaw)
- Convulsions
- Extreme sensitivity to touch
- Anxiety
- Excessive salivation
- Rigid extension of the legs (front legs extended forward and together, back legs extended backwards and together)
- Death

Treatment:
- Tetanus antitoxin
- Antibiotics
- Wound treatment

- Electrolytes and plentiful water

Prevention:
- Vaccination (CD&T)
- Cleanliness of bedding, stalls and paddocks
- Avoid overcrowding

The vaccine for Tetanus is commonly combined in a three-way injection, which provides protection for Clostridium perfringens (types C and D), and Tetanus (T) (see Enterotoxemia Vaccine Schedule, page 86).

Tetanus is much easier to prevent than it is to treat. High doses of antibiotics and tetanus toxoid will be required, and even with that, a veterinarian may need to step in to give fluids and medications intravenously.

The incubation period for tetanus is 10 to 20 days; however, once symptoms appear death quickly follows in 24-36 hours.

Clostridium tetani commonly contaminates both soil and animal feces. The disease is normally contracted through wounds, disbudding, castration, difficult births and surgical procedures.

***

Toxoplasmosis
Toxoplasmosis is a disease caused by the protozoan parasite Toxoplasma gondii.

Clinical Symptoms:
- Death and resorbed fetus (early gestation)
- Abortion, still births, birth defects, mummification of fetuses, and weak kids (mid to late gestation)

- Kids born with no outward signs of disease, but are carriers (late gestation)

Treatment:
- No known cure or treatment

Prevention:
- Coccidiostat added to feeds
- Avoid feed, hay, or grasses contaminated with cat feces
- Careful removal of aborted fetuses and birthing materials (with **gloved hands, humans can contract toxoplasmosis**)
- Control cat population on farm
- Cover feeds

Notes and Tips:
As of this writing, a vaccine for Toxoplasmosis is not available in the United States. There is a vaccine available in other countries for sheep. Continue to ask your veterinarian for updates on available vaccines.

Toxoplasmosis is dangerous to many warm-blooded animals – birds, sheep, goats, cattle, pigs and poultry, and cats are a carrier. Toxoplasma gondii sheds into the environment through feces. Cats, especially kittens less than 6 months old, defecate in straw, hay, and some feeds. For this reason, discard feeds contaminated with cat feces, or the feces from any animal.

Toxoplasmosis is a disease of pregnancy. Signs of the infection rarely make a presence in a non-pregnant animal. In pregnant goats, the outcome rests entirely on when the doe was exposed. The parasite multiplies in other organs, especially the brain and muscles, and then it invades the placenta and fetus. Exposure during the first half of pregnancy results in resorbed fetuses and abortions. In the second half of pregnancy, weak kids, birth defects, stillbirth and mummified kids occur. Contraction of the disease later in pregnancy results in kids born with no outward signs of the disease, but they are carriers.

Toxoplasmosis normally affects a doe's first pregnancy, and generally does not affect subsequent births.

Bucks also carry Toxoplasmosis, with no outward clinical symptoms.

Humans can contract Toxoplasmosis. It is especially dangerous to pregnant women and their unborn. Wear gloves when removing aborted fetuses and birthing materials. Raw goat milk and goat meat can also carry the pathogens.

***

Urinary Calculi (Water Belly)

This disease occurs when crystals (stones), form in the urinary system and lodge in the urinary tract. The cause of this disease is a metabolic imbalance, often initiated from improper feeds and a lack of drinking water.

Clinical Symptoms:
- Difficulty with urination
- Restlessness
- Hunched up body posture
- Bloated belly appearance
- Teeth grinding
- Holding stretched out appearance for lengthy periods of time while urinating
- Little to no urine coming out
- Kicking at belly

- Death (occurs rapidly)

Treatment:
- Contact veterinarian

Prevention:
- Avoid feeding bucks and wethers a high percentage of concentrated feeds (grain)
- Ensure concentrated feeds are correctly balanced for goats (proper calcium to phosphorus ratio, 2:1)
- Provide plentiful clean fresh water
- Provide plentiful roughage in the diet
- Include Ammonium Chloride in wether and buck feed
- Avoid the feeding of corn based feeds (contains a high percentage of phosphorus)
- Castrate males after they are 30 days of age (see page 195)

Notes and Tips:
Bucks, and especially wethers, are prone to this disease. Does have a longer and much straighter urinary tract, which allows solids to pass.

As a measure of prevention, do not band (castrate) a buckling until he is at least 30 days old. Castration causes hormonal changes, which can hamper growth of the urethra.

Urinary Calculi results in extreme pain and death. Contact your veterinarian as soon as you suspect urinary problems. Treatment of a male goat normally includes surgery, but may include infusing the urethra with a solution, or the injection of a tranquilizer to relax the muscles of the urethra, which helps the solids to pass.

***

White Muscle Disease

White Muscle Disease is a degenerative disease that affects skeletal and cardiac muscles. It is caused by a deficiency of selenium and/or vitamin E.

Symptoms:
- Mild stiffness to obvious pain upon walking
- Inability to stand (and trembling in pain)
- Stiff gait and hunched appearance
- Normal appetite (lessens when the animal becomes weak)
- Symptoms appear to worsen after exercise
- Newborns unable to rise
- Poor coat
- Lack of milk
- Difficulty breathing (heart muscle symptom)
- Frothy or blood stained nasal discharge (heart muscle symptom)

Treatment:
- Vitamin E and selenium injection (veterinarian prescribed)

Prevention:
- Quality feeds
- Avoid overstoring of grains and feeds that contain vitamin E
- Loose minerals that contain both selenium and vitamin E (along with other minerals – including copper, specifically blended for goats)

- Annual selenium/vitamin E injection (veterinarian prescribed)

Notes and Tips:

Selenium (Se) deficiency is associated with selenium deficient soils and feeds grown on these soils. Your veterinarian will help you determine local and regional selenium deficient areas, and will prescribe necessary treatment.

Vitamin E deficiency is a reflection of feed quality, not so much related to the soil type. Fresh pasture and legumes are a good source of vitamin E. Prolonged storage will result in the rapid loss of quality, up to 50% per month.

White muscle disease is more common in newborns and rapidly growing animals.

Deficiencies that affect the skeletal muscles are easier to treat. When the heart muscles are affected permanent damage can occur which cannot be reversed.

When injecting a goat with a selenium supplement, follow your veterinary instructions carefully. Too much selenium can be toxic! Toxicity does not normally occur from feeds containing selenium, but toxicity does occur from injections. Too much selenium results in the same symptoms as a lack of, and toxicity cannot be reversed.

***

Internal Parasites

There are two major types of parasites, external and internal. In prior sections of this chapter, I discuss external parasites – Mange and Lice, and one internal parasite – Coccidia. In this section, I cover internal parasites, specifically worms, which consist of two types – gastrointestinal (stomach worms), and those that affect other organs (liver, lungs and brain).

A goat owner will spend more time and money on the treatment and prevention of internal parasites than they will any other affliction or infestation.

Learning about the origination of the larvae and worms, the life cycle of each of these parasites, and knowing what symptoms to watch for, lends miles towards the goat owner's diagnosis and swift treatment. Calling a goat "wormy," and treating blindly, can be costly and may shorten the life of a goat.

Gastrointestinal Parasites

Goats always have a small number of parasites in their intestines, and they seem to form immunity to many types of worms, however, one species of roundworm is more prevalent and much more deadly than any of the others – Haemonchus contortus, the Barber Pole worm.

Barber Pole Worm (Haemonchus contortus)

The Barber Pole Worm is the most common and deadliest internal parasite known to goats.

This parasite imbeds itself and feeds on blood in the second stomach, the abomasum. It multiplies rapidly, which is why anemia is usually the first outward sign of an infestation.

Clinical Symptoms:
- Pale inner eyelids (anemia)
- Weight loss
- Tail down
- Diarrhea
- Rough dull coat
- Runny eyes

- Lowered milk yield
- Dehydration
- Lethargic
- Bottle Jaw (see page 73)
- Poor growth rate
- Reduced reproduction performance

Treatment:

- Dewormer, schedule:
  Day 1 and day 10 (up to 3 times, to kill adult worm and larvae as they mature)
- Withhold feed as much as possible prior to deworming (slows the passing of the drug down in the GI track)
- Probiotics (may be needed for rumen balance if infestation is substantial, especially if more than two dewormer treatments are given)
- Vitamin – iron – mineral feed supplement (see page 60)
- Vitamin B Complex, fortified with Thiamine (if symptoms include diarrhea)
- Electrolytes and plentiful water (if goat is dehydrated)

Prevention:

- Balanced nutrition and feeds or minerals that include copper
- Pasture rotation
- Never feed off of the ground
- Cleanliness in barn, bedding, water troughs and feeders
- Avoid overcrowding

Notes and Tips:

Female adult Barber Pole Worms have a distinctive red and white stripe that runs up their approximately 1" body, which resemble an old-fashioned barber pole. The male worm is solid red.

Barber Pole larvae thrive in moist, warm and humid conditions – which are summer conditions in many regions. After it rains, or early in the morning when pasture is damp with dew, the worm makes its way up from the dirt to the tip of a grass blade. As the grass dries under sunlight and heat, the larva retreats to the soil. During warm, humid and damp seasons, especially if you have small pastures, detain your goats until the upper levels of the blades are dry from the heat and the sun.

This worm is a proliferous egg-layer. Female Barber Pole Worms can lay several thousands of eggs per day, upwards of 5,000 to 10,000, with a life cycle of three weeks. 10,000 of these adults can kill a goat. This gives you an idea of how quickly an infestation can drain the health and life out of a healthy goat. A goat can appear healthy one day, and become anemic, weak; unable to stand on all four hooves the next.

I have found that the clear dewormers are best for the Barber Pole Worm. As of this writing, several dewormers are no longer effective (especially those that are white colored), and the loss of effectiveness appears to be an ongoing problem. Your veterinarian will determine the best dewormer to administer based on fecal samples from your animals.

***

Tapeworm (Moniezia expansa)

Tapeworms are common to goats, but unlike the round or Barber Pole Worm, clinical symptoms rarely make an appearance.

Clinical Symptoms:

- Weight loss
- Poor health (in severe infestations)

Treatment:

- Dewormer

Prevention:

- Balanced nutrition and feeds or minerals that contain copper
- Pasture rotation
- Never feed off of the ground
- Cleanliness in barn, bedding, water troughs and feeders
- Avoid overcrowding

Notes and Tips:

The tapeworm is flat, ribbon-shaped, and lives in the intestines of animals. The adult tapeworm sheds egg-containing segments in feces. The shear fact that farmers can see the worm segments causes much concern, however, these worms do not reproduce in huge numbers and they do not quickly suck the life out of a goat, as do the Barber Pole Worm.

The eggs of the adult tapeworm are visible microscopically; however, the adult worms are not.

When a major tapeworm infestation is diagnosed or suspected, consult with the veterinarian for deworming procedures. Intestinal blockage can occur when a large number of tapeworms are killed at once.

Notes and Tips (all species of worms):
Prevention

Prevention is the key to avoiding worm overload in a goat. This starts with sanitation – clean water troughs, feeders and feed free of fecal material, cleanliness in the barn and sleeping areas, pasture rotation, and no overcrowding.

Adult goats should be dewormed as often as needed to control the various types of worms. Nutrition should be optimal, but even at that, culling a line of gene-related goats from the herd may be necessary. Some hereditary lines, breeds and cross breeds, are more parasite resistant than others.

If you have the land available, consider that goats are browsers by nature. When goats have browse to select from, they will not be ingesting as many worms. Worms come from the ground, not plants and trees that goats reach up or into.

Consider alternating paddocks with other livestock that do not share the same type of worms. For example, alternate the paddock with a horse, or share the areas if they do not kick or chase after the goats.

Worm infestations, such as the Barber Pole, that thrive in damp and warm conditions, can occur at any time, but are more seasonal. The larvae are nearly dormant in winter months. Always be prepared for an outbreak after goats return to pasture following dampness, before the sun heats the pasture sufficiently. Consider feeding hay when warm and damp weather hangs around for extended periods.

Rotation of pasture and paddocks, along with the avoidance of overcrowding are solid ways to avoid parasite infestation.

Fecal Counts

Fecal counts, the process of microscopically counting worm eggs contained in manure, is the ultimate way to detect and manage the worm population in your herd. Fecal counts can be performed by a veterinarian or laboratory, but I have to warn you that the fees can be costly, especially when testing is frequent, and when testing a large herd. I suggest selecting a number of goats randomly for each test, around 10% of the herd. Doing so will give the results of the average overall worm population in your herd. Select individual goats for testing when they have had problems with worm overload in the past, or when you suspect that they currently have a problem with worms.

An alternative that will save you money, and will provide instant results, is to purchase a microscope and test materials and either seek lessons from a knowledgeable friend, laboratory or veterinary technician, or study the technique yourself (i.e. online or books), which is exactly what I do. I purchased an inexpensive microscope and equipment: flotation solution, spoon or stirrer, straining device, test tube, microscope slides and cover slips, and I studied photographs and characteristics of worm eggs. The fecal count, by the way, is also a way to diagnose and monitor Coccidiosis (see page 80).

Fecal counts combined with regularly checking the color of the lower eyelid, helps with determining which goats needed to be dewormed. FAMACHA is a method, used as a tool by goat farmers, developed by a group of veterinarians and scientists in South Africa and, and validated in the U.S. by members of the Southern Consortium for Small Ruminant Parasite Control (www.scsrpc.org). FAMACHA involves examining the color of the lower eyelid, matching the color on a chart that ranges from red to almost white (see Anemia, page 67). The lighter the color of the lower eyelid, the more anemic the goat is.

It is important to know, do not depend entirely upon the color of the eyelids when looking for worm infestation. Some worms are not blood sucking, and some goats may not appear ill, at all. They may even have bright pink eyelid color until the day they show outward signs of infestation. Fecal testing, the FAMACHA method, combined with sanitation practices, are all important pieces in the prevention of worm infestation.

Deworming Kid Goats
Younger goats, from birth to several months old, have not built up resistance and tend to be more susceptible to internal parasites. Schedule dewormings for young goats at 3 - 4 weeks old, and at weaning when they are most susceptible (2 ½ - 3 months old).

Deworming During Pregnancy and After Birthing
Practice caution when deworming pregnant goats. Some dewormers are not safe during gestation. Read the label carefully. If possible, unless the goat has a severe parasite infestation, consider holding treatment off until after she gives birth.

When a goat gives birth, hormones are a catalyst to worm infestation. I recommend worming the dam the day she gives birth, followed by a second worming in 10 days.

About Dewormers

Worms have morphed and have become resistant to many commercial dewormers (especially the white liquids and pastes).

As of this writing, these are the types of dewormers available:

- Benzimazoles – white – resistance found on most farms
- Macrocyclic Lactones – clear – resistance growing, but still effective in many regions
- Levamisoles – yellow – resistance growing

There is debate among the goat farming communities about the effectiveness of rotating the dewormer type. Some feel rotation leads to immunity to the dewormer. I always keep several different dewormer types in the goat medicine chest. I use clear commercial dewormers, but I switch between types on a resistant goat. I recommended doing what is best for your herd, what works best to keep the worm population count at its lowest, and if your process works, do not change it.

Most dewormers are not labeled for goats. Check with your veterinarian for exact dosages and correct application. Goats have a unique metabolic system, which makes oral treatment the best option (even when the bottle is labeled injectable or otherwise).

In some regions, a purple Macrocyclic Lactone type dewormer is used on goats. It is labeled as a pour on treatment, but is given orally. Check with your veterinarian for safety and dosage information.

When drenching, i.e. dosing with a liquid dewormer, do not overdose. If you can afford a goat scale, it helps with giving the correct dosage by weight.

***

Worms (non-gastrointestinal)

Liver Fluke

The Liver Fluke begins its life cycle through the shedding of feces. An egg develops by hatching and it migrates to a host, which is a snail. The Liver Fluke then multiplies and develops into an infective form, calcarea, which moves to pasture once the outside temperature and moisture level is suitable for survival. The goat ingests the calcarea, which migrates through the stomach, the intestinal wall, and on to the liver. It grows inside the liver, destroys tissue, and eventually finds a bile duct where it produces eggs. This initial infestation process takes place within 6-8 weeks, however, extreme damage and outward signs of infestation are slow to emerge.

Clinical Symptoms:
- Weight loss
- Decreased fertility
- Rough coat
- Anemia
- Reduced milk production
- Rapid heartbeat (advanced stage)
- Sub-normal temperature (advanced stage)
- Distended stomach (advanced stage)
- Feces dark black (appear shriveled and pointed on end, advanced stage)

Treatment:
- Dewormer (see veterinarian for testing and treatment)
- Vitamin – iron – mineral feed supplement (see page 60)

Prevention:
- Avoid pasture that remains wet or has poor drainage – especially low lying areas in warmer climates that hold water
- Allow grazing after the sun heats the pasture which causes snails to retreat
- Quarantine new animals to the farm for at least 30 days
- Quarantine animals with suspected cases for at least 30 days

Notes and Tips:
Fecal tests for Liver Fluke must be completed by a veterinarian. It cannot be detected through simple methods common to a goat owner. The veterinarian will base treatment upon age of the fluke and condition of the goat. Veterinarians are aware that Liver Fluke and the Barber Pole Worm look very similar under a microscope. In areas prone to Liver Fluke, use a dewormer for multiple types of parasites.

As of this writing, there is one dewormer on the market that is labeled for Liver Fluke; it contains ivermectin 10% and clorsulon. In regions where snails are common, consult with your veterinarian for use of clorsulon as a single preventive measure.

Quarantine of infected goats is extremely important because the eggs are shed through feces for 3 to 4 weeks after the adult Liver Flukes are killed through treatment.

Because symptoms of Liver Fluke closely resemble those of round worms, goats are treated with dewormers that do not work. The health of the goat continues to go downhill, which leaves the farmer thinking the animal has built resistance to the dewormer. When Liver Fluke is finally diagnosed, the disease may have already caused serious damage.

**

Lungworm

Lungworms are roundworms that infect the lungs and/or bronchial tissues of goats. There are two types: Dictyocaulus filarial and Muellerius capillaris.

Muellerius capillaris is a very tiny worm that imbeds itself in the lung tissue and generally does not cause severe problems. Dictyocaulus filarial is a larger worm that matures in the air passages and can lead to scarring and lung related diseases. Treatment is the same for both types of lungworms.

Lungworms use snails and slugs as hosts, while some types develop on plants in the pasture. The goat ingests the snail, slug, or larvae. The lungworm matures and lays eggs in the lungs or bronchial tissues. The eggs hatch into larvae, which are coughed up and swallowed. Then they pass through the goat's body and continue to contaminate the soil and pasture through feces.

Clinical Symptoms:
- Coughing
- Difficulty breathing
- Loss of appetite
- Pneumonia
- Bronchitis

Treatment:
- Levamisole (contact veterinarian for treatment and dosage)
- Repeat treatment 35 days from day one to ensure adult and larvae are killed.

Prevention:
- Avoid pasture that remains wet or has poor drainage – especially low lying areas in warmer regions that hold water
- Allow grazing after the sun has come out – causing snails and slugs to retreat
- Quarantine infected animals

Notes and Tips:
Lungworms are white and thin in appearance and can be up to three inches in length. The life cycle ranges from five to ten weeks, and remains infective to the goat for at least one year.

A goat with mild infestation may appear healthy. A goat with severe infestation will cough and have trouble breathing.

Young kids are particularly prone to pneumonia and bronchitis, secondary diseases caused by Lungworm infestation.

Treat for stomach or intestinal worms on day one. Treat for Lungworms on day two. This gives the worms room to move and lessens the chance for lung damage.

Animals continue to shed Lungworm larvae up to 20 days following treatment. Quarantine infected animals for three weeks.

Do not confuse Lungworm with allergy symptoms from dust, pollen or hay, which produces a dry cough, with clear discharge from the nose.

Fecal testing is not reliable for Lungworm. The eggs tend to be heavier than other worm eggs, which causes them to sink and drop below the microscopic field of sight.

Contact your veterinarian for available treatments and dosages. Levamisole can cause abortion in pregnant animals, and its availability may be limited (or not available at all) due to misuses in illegal drug making.

***

Meningial Worm (Deerworm, Brainworm)

The meningeal parasite, Parelaphostrongula tenius, is common in areas where white tail deer are abundant.

The parasite completes its life cycle in deer without harm to the animal. Deer shed the parasite through feces. Snails or slugs become intermediate hosts, which are ingested once again through grazing.

In goats, the Meningeal Worm larvae migrate into the brain and/or spinal cord and cause neurological problems within 10 – 14 days after the animal eats the snail or slug.

Clinical Symptoms:

- Limp or weakness in one or more legs
- Partially or completely paralyzed (severe infection)
- Blindness, head tilt, circling, lack of interest or inability to eat (severe infection in brain)
- Seizures and death (severe infection – difficult to treat at this stage)

Treatment:
- Anthelmintic (deworm in large and repetitive doses)
- Steroids (to reduce inflammation, see veterinarian for treatment)

Prevention:
- Avoid pasturing goats in areas with high deer populations
- Limit grazing in areas close to woodlands that may be infected
- Avoid low lying areas with damp vegetation prone to snails and slugs
- Avoid pasture that remains wet or has poor drainage

Notes and Tips:
As of this writing, a Meningeal Worm infection cannot be diagnosed in a live animal. Fecal counts are not accurate for this parasite because the parasite enters the brain and spinal cord and no indications of the parasite are shed through feces.

Successful treatment of this parasite greatly depends upon the severity of the symptoms, and the number of larvae ingested. If the goat can still stand upright, the chances of recovery are much stronger. Improvement can occur upon treatment, but total recovery of this parasite can take weeks to month.

Additional therapy may be required. This may include soft bedding, putting the goat in a sling, and helping them to remain on feet.

Deworming on a regular schedule will help to prevent meningeal worm infestation, but it will also add to the likelihood of resistance to stomach worm treatments. It is best to prevent meningeal worm through the methods mentioned above; pasture in dry areas away from any contact with deer.

***

Poisoning

Grain overload and plant toxicity are the most common poisonings that goat farmers will encounter.

Treatment depends entirely upon the type of poison, and the severity – long term (chronic) or recent ingestion, affecting the animal. To determine the cause of poisoning, establish what the goats were exposed to (types of feed, pasture, weeds, and treats that may have contained sprays or chemical poisons).

Starvation is a forerunner to poisoning. The availability of plentiful pasture to graze, or quality hay, greatly reduces the chance that an animal will resort to eating weeds or other things that they might not normally seek.

Common Poisons:

Algae (in water)

Algae can make a goat ill, or can be deadly when the animal drinks from a pool of stagnated water filled with blue/green algae. Keep water troughs clean, filled with fresh water, out of direct sunlight during hot seasons.

***

Arsenic

Arsenic is a poison used to control rodents, and has been used in dips to kill lice and other parasites.

Clinical Symptoms:
- Sudden distress
- Severe abdominal pain
- Restlessness
- Groaning
- Increased respiratory rate
- Increased salivation
- Grinding of teeth
- Diarrhea
- Vomiting

Treatment:
- Contact veterinarian for treatment

Prevention:
- Discard old and unused medications and chemicals. If the products are needed on the farm, store and secure the solutions in sealed containers in an area not accessible to animals or children.

Notes and Tips:
Contact your veterinarian for treatment. Severe and sudden arsenic poisoning is difficult to treat.

When an animal ingests arsenic in small amounts over a long span of time symptoms may appear gradually. The initial symptoms are: loss of appetite, loss of hair, loss of condition and weakness.

Oral electrolytes are necessary when diarrhea sets in. In severe cases, a veterinarian may need to administer electrolytes intravenously to treat dehydration.

Animals that survive severe arsenic poisoning may end up with chromosome, bone marrow and renal damage.

***

Farm chemicals (Organophosphates)

Organophosphates are commonly used as crop sprays, and are applied externally for lice and tick control.

Clinical Symptoms:

- Trembling
- Muscular twitching
- Stiffness in walking
- Increased salivation
- Diarrhea
- Breathlessness
- Pinpoint pupils
- Convulsions

Treatment:

- Contract veterinarian for treatment

Prevention:

- Discard old and unused medications and chemicals. If the products are needed on the farm, store and secure them in an area inaccessible to animals and children.

***

Grain

When people talk about grain poisoning, they are referring to either overfeeding, or a quick change over to a different type of grain, or the feeding of spoiled feeds or moldy hay that can lead to the release of toxins in an animal's body. The resulting diseases are Enterotoxemia (see page 85) and Listeriosis (see page 99).

***

Lead

Lead is the most common type of poisoning in farm animals. There is less lead in our environment today because it is no longer used in paint, and is carefully monitored in the production of other materials, however, items that may contain lead are car batteries, lead shot, and sump oil. An older barn with flaking paint may lead to contamination, and it is especially dangerous to animals that like to mouth painted surfaces.

Clinical Symptoms:
- Depression
- Diarrhea
- Abdominal Pain
- Loss of balance

Treatment:
- Contact veterinarian for treatment

Prevention:
- Cleanliness of pasture and paddocks (remove farm equipment, such as oils, old batteries, tin cans, etc.)
- Paint over or remove painted materials that may chip or flake. Note, lead is dangerous to humans. Do not attempt to sand, scrape or remove paint from lead-based surfaces without contacting your local health department. In some areas, the procedure is regulated extensively, requiring the hiring of licensed professionals for removal.

***

Nitrite

Some plants are naturally high in nitrates, such as: pigweed, milk thistle, and poison hemlock. Crops, such as rye and barley, are not high in nitrates until certain conditions (drought, uneven distribution of rain, low light intensity) occur.

Clinical Symptoms:
- Depression
- Difficulty breathing
- Abdominal pain
- Diarrhea
- Muscle tremor
- Weakness
- Staggering
- Extreme nervous reactions
- Death

Treatment:
- Contact veterinarian for treatment

Prevention:
- Ensure pasture and paddocks are free of weeds and crops that contain nitrates

***

Rat and Snail Baits (containing Dicoumarin and Brodifacoum)

Rat and snail baits are deadly to animals, yet, goats are particularly drawn to these products, especially bran-based pellets. The chemicals in these baits contain anti-coagulants, leading to internal hemorrhaging after ingestion.

Clinical Symptoms:
- Drowsiness
- Loss of appetite
- Constipation
- Abdominal pain
- Diarrhea
- Death

Treatment:
- Contact veterinarian for treatment

Prevention:
- Avoid the use of rat or snail bait in any area within the reach of animals or children
- Avoid feeding spoiled hay or feeds

Notes and Tips:
Dicoumarol (dicourmarin), the active chemical in rodent bait, is a toxic chemical derived from coumarin, which is also produced by the heating and spoilage of hay or silage. Ingestion leads to toxic diseases, such as Listeriosis (see page 99).

***

Selenium
Selenium is highly toxic to goats when given in greater doses than the body requires. When administering selenium through injection, carefully follow veterinarian instructions.

***

Poisonous Plants

The following are some of the common plants, shrubs and trees known, and suspected to be, poisonous to goats:

Allspice
Avocado
Azalea
Black Locust
Black Snake Root
Boobialla
Boxwood
Bulb plants (all)
Camphor Laurel
Elderberry
Elm
Fir
Gardenia
Green Cestrum
Hemlock
Hemp
Hibiscus
Kale
Laburnum
Lantana
Laurel
Lilac
Lily of the Valley
Lupines
Macadamia
Magnolia
Marijuana
Milkweed
Monkweed
Mountain Laurel
Nightshade

Noogoora Burr
Oleander
Plum
Poison Peach
Pokeweed
Poppies
Potato
Privet
Rhododendron
Rhubarb
Rhus
Sally wattle
Stagger Grass
Sudan Grass
Tomato Plants
Variegated Thistle
Wild Hydrangea
Wild Parsnip
Yew
Choke Cherry

Clinical Symptoms:

- Symptoms widely vary by the type of poisonous plant ingested (affecting various organs and nervous systems – including the heart, liver, lungs, stomach and brain)

Treatment:

- Contact veterinarian for treatment

Prevention:

- Inspect new pasture before turning goats out to graze
- Do not feed unidentified greenery (cuttings or dried leaves)

- Avoid chemically treated areas (areas treated with herbicides)
- Ensure goats cannot reach through or over fencing to poisonous plants

Notes and Tips:
A list containing all of the plants that are poisonous to goats would encompass many pages of this book. I have only included the most common poisonous plants to goats that are found in many parts of the United States. I recommend contacting your veterinarian and local livestock extension office for a complete list of poisonous plants common to your area.

As a general rule of thumb, many yard or ornamental plants are highly toxic to goats.

Some toxic plants taste good to goats, and some are noxious enough for animals to avoid altogether. The toxicity of plants is greatly affected by weather – some are more toxic after they wilt, and some are more toxic following rapid growth and after rain.

Plants treated with herbicides tend to taste better to goats. Remove animals from these areas for up to 2 weeks after treatment, or do not allow them access at all.

***

Notes and Tips (All Poisons):
Immediately contact your veterinarian when poisoning is suspected. The vet will supply you with anti-toxins and detailed instructions.

In some situations, the veterinarian may tell you to wait it out, especially in cases where treatment may not help (example: a poison that affects the heart). In some cases, the vet may ask that you keep the goat calm, and may ask you to encourage them to eat and drink water. In these cases, as much green fresh green feed you can get your hands on will be helpful such as: cuttings from non-poisonous trees (the goat will look at this as a treat and may be enticed to eat), and top quality fresh hay.

The following is a basic list of items to keep on your farm for poisoning treatment:
Injectable C&D anti-toxin
Tetanus anti-toxin
Magnesium hydroxide
Activated charcoal
Mineral oil
Adult goat stomach tube and mouthpiece
60 cc kid syringe and stomach tube
Prescription pain medication and inflammatory

Chapter 9
Coat, Hoof and Horn Care

Coat Care

Goat coats are generally maintenance free, meaning, no combing, trimming, or special management is required.

The coat reflects the health of the goat. A smooth and supple coat is an indication that the goat is free of internal and external parasites. A dull coat, one that looks shaggy, with spotty growth patterns, or bare patches, is indicative of poor health or parasitic overload.

With the exception of fiber goats, the Angora breed, shearing is not necessary. You may be tempted to shear a goat during extremely hot weather, but I recommend not doing so. The coat thickens and protects the goat from heat loss in cold weather, and the summer coat thins down, which combined with natural skin oils protects the goat from sunburn and excess heat.

Animals are often sheared for goat shows. In colder climates, or when the goats are in stalls at an indoor air-conditioned show, consider manmade coats to prevent chilling. To avoid sunburn, limit the sheared goat's exposure to direct sunlight.

Hoof Care

Goat hooves require periodic maintenance. Trimming helps to prevent diseases of the hoof, and helps the goat to walk and stand properly, which affects so many aspects of their lives – feeding, milking, exercise, and breeding.

There is no set schedule for hoof trimming, however, plan to trim about four times a year. Some hooves seem to never need trimming, while others grow rapidly. Check the hooves regularly for excess growth, especially the back hooves, which do not wear down as readily as the front.

Begin trimming the hooves of young goats as early as the need arises. This helps the kid acclimate to handling. It is much easier to trim the hooves of a tame goat, one that is used to your touch, than it is to work with a full-grown kicking and bucking goat with overgrown and impossibly thick hooves.

The easiest way to restrain a goat for hoof trimming is to place them on a milk stand or stanchion which holds their head in place. By doing so, the goat will be standing on a flat surface 12" to 18" off the ground, which helps you to see what you are doing, and it saves your back muscles from excess strain. You may even be able to sit down during part of the trimming process.

Materials to have on hand for hoof trimming: rubber gloves (protects your hands from cuts and blisters), sharp hoof trimmers (see Suppliers, page 241), soapy water and a brush, disinfectant, and blood stop powder

Ensure your goats have had their annual Tetanus vaccination before trimming hooves.

How to trim hooves:

- Clean dirt or debris from the sole and from between the toes (use the trimmer or a hoof pick)
- Trim between the hooves, where the heels meet – taking care to not trim too deep
- Trim the sidewalls of the hoof. The desired result is hooves that are even and flat with the sole of the foot.
- Trim the sole if needed, but do not dig into the pink. Bleeding can and will occur.
- While trimming, watch for separation of the wall of the hoof from the hoof itself. You may need to cut dirty and damaged areas out to eliminate disease (see Foot Rot and Foot Scald, page 147).
- Trim just a little at a time to avoid cutting too deep. Keep in mind, the thickness of the hoof is different in every goat.
- When hooves are overgrown, or when disease is present, trim only as much as needed. Trim in small increments and prepare to trim again within a few weeks.

Special Notes - Hoof Trimming

Know what a perfect hoof looks like by studying the hooves of young goats. Observe the area between the toes and notice how the young goat stands. This will give you an idea of how you want trimmed hooves to look, as well as how the goat's stance looks on well-trimmed hooves.

Take your time and always trim gradually. Do not cut into or expose pink areas of the hoof. If you do cut too deep, immediately disinfect the area and apply blood stop powder if blood appears.

If you are new to hoof trimming, I recommend finding a goat owner that is willing to give you a demonstration. One-on-one training is best, an online video may also help, as well as online tutorials.

Horn Care and Removal
The burning question seems to be, horns or no horns?

Cons to Horns
Many dairy goat owners disbud (burn off the horn buds) as soon as the kids are strong enough for the procedure. They do this to protect the dam's udder from injury.

Owners that show dairy goats have the horns removed because dairy goats are disqualified from showing if they have horns. Some clubs, such as local 4-H, require disbudded or hornless animals in the show ring; however, check with your local branch because these rules vary.

There are additional cons to consider. A goat with horns will invariably get their head stuck in a fence. This can leave a goat exposed to attack from predators, or may lead to injury.

One concern that regularly pops up is whether to paddock horned and hornless goats together. I have kept goats with and without horns in the same area, and neither has suffered injuries directly related to horns.

Pros to Horns
Horns do serve a purpose. They protect the skull, and they protect the animal from predators if the need to fight for life occurs. Horns also function as body temperature regulators.

Disbudding

The most common method of horn removal is disbudding, which consists of burning off the horn buds before they actually form into horns. Disbudding is done when a male kid is 3 days old, and a female is 4 days old. Disbudding can be done at a later age; however, scurs commonly develop, especially in male goats. Scurs are unsightly partial horns that grow from any remaining portion of a horn bud. Male kids have a larger horn or bud base than female kids. The earlier the disbudding takes place, the smaller the horn bud base.

Dehorning

Dehorning is the method of removing grown horns from mature goats, only a veterinarian should do this. A small part of the skull is removed in the process.

Grown horns can also be cut, but it is a dangerous procedure that can cause pain and bleeding.

Do not attempt either of these procedures (disbudding or dehorning) on your own. A veterinarian will remove horns in a humane manner, with the least amount of pain – with the addition of anesthesia and pain medications, and with little or no blood loss.

Keep in mind, most goat owners do not dehorn after the animal's horns have grown beyond buds.

Polled

Genetic lines of animals that do not grow horns are called polled goats.

Polled goats are a great option for owners that do not want to deal with disbudding, but it is important to note that a polled animal bred to another polled animal increases the chance of hermaphrodite offspring. Hermaphrodite goats have both male and female sex organs, making breeding difficult or impossible, and offspring are often un-thrifty animals. Male goats that are not hermaphrodite, but those that possess the hermaphroditism gene, may have blocked semen, with an inability to breed. Female polled goats are generally fertile. Polled female goats are an option for a goat owner that does not plan to breed back to male offspring.

Breed a polled doe to a buck with horns, one that is not from the same genetic line, normally results in a 50/50 chance of polled offspring.

Horns and Fences

If you experience continual problems with a horned goat getting their head stuck in a fence, there are ingenious ways to resolve the problem. Tie or tape a length of 1" PVC or wooden dowel horizontally between the tips of the horns. Do not drill holes into the horns. Make sure the PVC fits without being overly tight, which could cause the horns to grow in an unnatural direction! This method allows the horns to grow, but prevents the goat from sticking their head through small openings. It also trains the goat to stop making the attempt.

To prevent the goat from tangling in brush or tree branches, do not attach the PVC to the horns in a manner where it extends beyond the outside of the existing horns.

Our Experience with Dry Skin

Some goats have dry skin during colder weather, and this occurs more often with goats that are proliferous milkers. We check for parasites, and then we go to work on the skin. We provide supplemental sources of nutrition, such as organic black oil sunflower seeds

(BOSS), which is high in vitamin E, zinc, iron and selenium, and we brush the coat regularly with a soft rubber curry comb to distribute skin oils throughout the coat and skin.

Our Experience with Polled Goats

We raised beautiful Spanish/Boer cross does for years. They often produced polled bucks and does. We did not breed the related male offspring back to the does, and we never had a hermaphrodite kid born.

Twins born to our polled dams commonly resulted in one polled and one with horns.

Other than dairy goats, whose heads are locked in place on a stanchion while being milked, which makes disbudding desirable, we do not have a horned or hornless preference.

Chapter 10
Breeding and Pregnancy

Know Your Goals

Will you breed your goats, or not? This is one of the most important decisions that a future goat owner must make. Knowing the breeding goals of your herd will affect how you lay out fencing, housing, and whether you castrate, sell, or keep male offspring.

Are you able to devote time to your animals? Do you have the flexibility to tend to goats in labor (at all hours of the day or night), or to the occasional weak or ill newborn?

Do you plan to milk goats? Breeding is a requirement for milking.

If you do not plan to breed, if you plan to own goats as pets, consider purchasing does that are not bred, or purchase wethers. Consider not purchasing bucks until, or if, you decide to breed.

Before You Breed

The health of your herd as a whole is of great importance to the success of your breeding season. Does should not be overly thin, nor should they carry excess fat at the time of breeding. Bucks, because they expend a great amount of energy during breeding season, should also be in good condition year round.

Remember, the buck is half of your herd, and without his good health, he does do not breed well.

Find a veterinarian that is knowledgeable about goats before you breed. I suggest doing this before you own goats, but especially prior to breeding in case you need assistance with birthing.

All animals should be free of disease, with healthy hooves, and with a clean environment prior to breeding.

Ensure your farm has an area in which to set up birthing pens; a dry area that provides shelter from rain and snow, extreme cold temperatures, cold and damp drafts, and predators.

Breeding Age

The age of the doe you plan to breed is of utmost importance. The most common initial breeding age is one year of age, however, the breeding age must also depend upon the does weight and maturity. If the doe is underweight, or experiencing slow growth, breeding at one year of age is not advisable. Keep in mind, a goat reaches full growth at approximately three years of age, but breeding before then does not affect the growth of the goat.

Do not breed a doe that is less than one year old, or prior to the time her body grows and matures sufficiently, which is when she has reached 60-70% of her adult weight. Some meat goats, especially the Boer breed, grow rapidly, and reach this weight (approximately 80 pounds for a Boer) as early as 8 months old. Does that are bred too early may have difficulty giving birth, which can lead to internal injury, difficulties with future births, and death while attempting birth.

The age of the buck is important as well. He obviously will not be the pregnant one, but he needs mature bone and muscle structure in order to maintain health during the breeding season.

Bucks and does can breed as early as 7 weeks old, which is obviously a very poor idea. Separate the young kids before they are mature enough to breed.

Breeding within Bloodlines

Goats within the same bloodline can be bred, which is called line breeding. Know that line breeding may increase negative characteristics in your herd, which includes health issues, poor bone structure, and the tendency within a line of genetics towards internal parasites. On the other hand, line breeding can multiply good qualities, such as motherly instincts, personality, milking abilities, and excellent growth patterns. Weigh out these options before breeding. If your decision is to breed within the same bloodline, do so between a doe and her father, or a buck and his mother, or another combination other than brother and sister.

Breeding Season

Some goat breeds, such as the Pygmy, are interested in breeding throughout the year. Other breeds, such as the Nubian or Boer, are seasonal breeders. Bucks can breed whenever the opportunity is in front of them. Does breed when they are in heat (in estrus), however, many breeds (see Chapter 4) tend to only go into season when the temperatures drop, and as the daylight hours shorten, which is very late summer and autumn in North America. These seasonal breeders continue breeding, but more sporadically, until January or February.

Intentional breeding, year round, of seasonal breeders is a tricky process. I enclosed a buck in a paddock with four dairy does for 45 days, off-season, which would have covered the estrus period for each doe twice, as well as a few extra days. No breeding occurred. Neither the buck, nor the does, showed any interest in breeding. This is not to say off-season breeding is impossible, I have had kids born four to five months later than the rest of the herd, but typically, it is difficult to break these hormonal and seasonal tendencies.

Estrus

When the doe comes into heat it is called estrus. Estrus lasts 12-48 hours, with an average length of 36 hours. Estrus occurs every 18 to 23 days during the breeding season, with an average cycle of 21 days.

Signs of estrus include flagging (wagging) the tail, swollen vulva, frequent urination, mucous discharge from the vulva, vocalization, attention turned towards a buck, mounting other does, a drop in milk production, and most importantly, if the doe is in the paddock with a buck, this is when the doe will allow the buck to mount her.

Bucks in Rut

Even though bucks can mate at any time, the breeding season occurs when hormones kick in and the bucks put on a show for the girls, called rut.

When a buck is in rut, he urinates on his front legs, face, and beard. He urinates into his mouth and he curls up his lips. The urine produces a musky odor that attracts female goats.

He may become difficult to handle when he is rut. Bucks in rut fight more aggressively amongst themselves, to the point they may need separation until rut season ends. This is a good reason to limit the number of bucks that you keep on your farm, however, more than one is good for cross breeding, and an extra covers the need for a replacement buck in case one goes ill.

Bucks are also very vocal during rut. Even the youngest of goats will make guttural noises. It is a humorous scene as they repeat, "woop...woop...woop," while they simultaneously paw or kick the ground with one hoof.

Bucks will also use their tongue to gain the attention of a doe. Do not be shocked if he does this at you too! He will either hang his tongue out the side of his mouth, or he will flap it while making a blubbering noise, and he can do all of this while he craftily paws the ground.

A young buck may not be as fertile as a mature buck. I recommend detaining him from breeding until he is 7 months old, and then, if you have a large herd, allow him to breed only a portion of the does.

I have had the surprise of color and characteristics pop up in offspring that should not have been. I later discovered that the buck hopped the fence overnight, did his mating ritual, and jumped back inside the buck pen each morning before I arrived to feed. Bucks are hard on fencing during breeding season, and they are determined!

You may not want to feed a buck year round. One option is to borrow a buck for the mating season, or you can transport your in-season doe(s) to a buck.

Artificial Insemination

Artificial insemination (AI) is also an option, especially when the purpose of the herd is for show, which necessitates the need for a line of excellent genetics and milking abilities. The options for purchasing semen are nearly unlimited by breed, pedigree (genetics), and cost.

Semen is collected from the buck and is stored in glass straws in nitrogen tanks. Later, the semen is thawed and inserted into the doe. Obviously, AI has benefits: precise calculation of the due date, the introduction of specific genetics into the offspring, and monetary savings by not having a buck to maintain and feed. On the other hand, AI can be an expensive process, injury to the doe can occur, and semen can be destroyed if not thawed properly. I recommend hiring a well-trained technician for this process.

The Act of Breeding
Do you have the picture yet of the strange behavior goats exhibit when they are charged up for mating? A goat's first love is eating, and their second love is mating.

The mating ritual resembles this: a doe in estrus (heat) will gain the attention of a buck by tail flagging, vocalization, or she will try to mount the buck. She will also urinate frequently, squat in front of the buck, and he will stick his nose and tongue under the stream of urine. He then curls his wet lips up so he can smell the scent. Sound lovely? The curling of the lips is called the flehman response or reaction, or flehming or fleming. By curling up their lips, while not breathing through their nostrils, pheromones and other odors are transfer directly into the vomeronasal organ. This is located in the roof of the mouth, with an opening just behind the front teeth.

The mating ritual continues, and I call this the dance. The buck will try to mount the doe and she will step aside. He tries again, and she may step aside again. She may even try to head butt him. The first time you see a buck mate a doe you will notice how quickly it goes. He mounts her, and moments later he is done. Generally, a doe will then hunch her back up, which a good indication that penetration has occurred.

Gestation

The gestation period, the span of time during which a goat is pregnant, is 143 to 155 days, with the average term of 150 days.

Similar to human pregnancy, a goat's gestation period consists of three trimesters. Each trimester, consisting of approximately 50 days, is critical to fetal development.

Goats typically birth in multiples. Twins are common, but are less likely in the first pregnancy. In subsequent births, triplets are not as common – but do occur, and much less frequently four kids are birthed.

Fetal Development

20 days	heart begins to beat
28-35 days	limb segments or buds appear
42 days	fetus is approximately 1-2 inches long; organs and body systems defined
42-49 days	formation of female mammary buds, male scrotal sac
49-56 days	ear canal opens, eyelids close
56-63 days	nostrils open

60 days	fetus is approximately 4 inches long
77-84 days	horn buds appear
90 days	approximately 10 inches long; hair starts to grow on forehead
98-105 day	teeth begin to erupt, hair on chest, lungs begin to form
126 days	hair will nearly cover entire body
140 days	body will be covered by dense hair
141 days	fetus is viable (can survive if born)

Pregnancy Detection

If a doe cycles into estrus 21 to 24 days after mating you may assume she is not pregnant, however, one may mimic a heat cycle early in pregnancy.

Blood testing for goat pregnancy is available. It is accurate after 35 days of gestation. A false positive for pregnancy occurs at about 5%.

Urine testing for pregnancy is also available and is accurate after the 50th day of pregnancy. These tests determine the estrogen levels in urine.

Ultrasound is another pregnancy detection method, with a 90% or greater accuracy rate after the 60th day of gestation. Ultrasound equipment can be purchased by the goat owner, but it is an expensive investment for those that are not into raising goats as a business. If you are in an area with many goat owners that are interested in obtaining an ultrasound, consider purchasing equipment as a co-op, sharing the cost of the equipment and supplies.

When a doe is within two weeks of giving birth, you may be able to see the movement of kids. If you see movement on the right side, you are likely seeing the functioning of the rumen, but if you see movement on the left side, especially if it looks like ripples, flutters or kicks, you may be seeing an unborn kid shifting around.

Another way to tell if a goat is pregnant, late in pregnancy, is the "bump test." Detain the goat on a stanchion or milk stand, or have someone hold her head. Stand behind and wrap your arms around her abdomen, and lace your fingers on her belly in front of her udder. Gently bump or push upwards with your hands. If kids are present, you may feel movement.

False Pregnancy

False pregnancies occasionally occur in goats

The doe goes into estrus; she is mounted by the buck, but due to nothing more than a hormonal glitch, conception does not happen. The doe goes into false pregnancy, and she produces estrogen levels, which give a false positive result from a urine test.

The heat cycles do not continue, and the doe's abdomen enlarges as fluids accumulate in the uterus. Her mammary glands (udder) enlarge, and she may produce milk. Her mothering instinct may also kick in as the false pregnancy reaches the average gestation term of 150 days. When this occurs, she may even appear to be looking for missing kid(s).

Some false pregnancies end within a few months. A full-term false pregnancy leads to the doe delivering fluids, called a "cloudburst," which consists of cloudy fluids, no placenta or fetus.

Generally, false pregnancies are not dangerous to the doe. Watch for signs of illness such as fever, lethargy, abnormal discharge, or depression. Watch the udder for signs of Mastitis (see page 102). In cases of false pregnancy, do not milk the goat. Doing so opens the teat orifice up which allows bacteria to enter, and milking can lead to a lengthier time before the goat returns to an estrus cycle.

Rarely, the goat will need veterinarian assistance, medications to help clear the fluids from the uterus.

Abortion and Embryonic Loss

Abortion, the loss of a fetus before it is full-term and viable, can occur at any time during a pregnancy, (see page 65).

Embryonic loss is common in goats. It occurs within the first four weeks of pregnancy due to a variety of causes. The fetus in this early stage is commonly absorbed. The goat returns to estrus cycles, and once bred, a viable full-term pregnancy normally follows.

When breeding to a related sire, and embryonic losses continue in a doe, switch to an unrelated buck.

Embryonic mortality is affected by body condition, extreme dietary changes early in pregnancy, extreme hot weather, genetic factors, hormone imbalances, age (too young or too old), disease, and too many fetuses in the uterine space.

Pregnancy Care

Feeding
Our biggest responsibility during a doe's pregnancy is to ensure she receives proper nutrition, which does not promote excessive weight gain or weight loss.

Throughout pregnancy, goats should be fed quality, dust and mold-free hay, preferably a legume type (such as alfalfa), which is mineral rich with calcium.

Throughout pregnancy, offer free-choice minerals formulated for goats, which always contains copper.

In the final 6 weeks of pregnancy, the period of time when kids are growing rapidly, offset the does nutritional needs with grain. A grain blend that contains 16% protein promotes healthy lactation. Begin feeding 1 cup of grain, or approximately a handful twice a day, and gradually increase this to 2 cups, twice a day, towards the final 2 weeks of pregnancy. This amount of grain is for medium to large does.

During the final trimester, consider feeding your dairy does on the milk stand. While they are eating, touch their udder, which prepares them for the upcoming milking process.

Milking Goats during Pregnancy
It is safe to milk pregnant does during gestation, however, dry them off (see page 224) after the 3rd month. This gives the doe 2 months to take in calories, and to absorb needed minerals, for herself and her growing kids.

Hoof Trimming

Hoof trimming should be completed prior to the final trimester. By doing so, you avoid over handling the doe, which equates to less stress. If the doe is accustomed to having her hooves trimmed while on the milk stand, and she can still balance on 3 hooves, there is no reason to avoid trimming.

Vaccinations

The final trimester is an optimal time to vaccinate a doe for Enterotoxemia (see page 85), Tetanus (see page 144), as well as supplement with selenium in deficient areas (see Notes and Tips, page 120). By giving these vaccinations and injections, you are providing initial disease protection to the unborn kids.

Housing

When the final trimester rolls around, you should have a plan in place for kidding pens. Goats tend to become defensive and possessive late in pregnancy. When this occurs, monitor to ensure the behavior does not become overly violent, which could cause injury or abortion.

Set up maternity pens before the 140th day of gestation arrives. Make sure the pen is clean, lined with fresh and dry bedding (I prefer the use of straw hay), free of mold because some goats will nibble on their bedding, ant and pest free, and in an area that is out of damp and cold drafts. Make sure there is enough room in the pen for the doe to stretch out, enough room for you if you are inside the pen while she is giving birth, and room for the newborn kids to move about as well. Keep in mind the doe and the kids will remain inside the pen for up to a week for bonding.

When setting up maternity pens, make sure the spaces in the wire are not large enough for small kids to climb through (because they will slip their way out!), and ensure there are no wire ends facing inwards, which can scratch an eye and cause injury.

The maternity pen must be roomy enough to allow for water buckets and feed pans. Hang water buckets high off the ground, or use shallow buckets to prevent drowning.

In cold climates, consider a box for the newborn kids to curl up in. Kids will snuggle against each other for warmth, and a box will prevent the dam from lying on the kids.

Mary L Humphrey

Chapter 11
Birthing

Preparing for Birth

Try to be present when the goat gives birth. Most goats are good mothers and problems seldom occur, but there is always the chance of a complicated birth.

Birthing Kit

Assemble a birthing kit prior to the final weeks of pregnancy.

- Phone number of experienced goat breeder (or veterinarian who you have already spoken with)
- Towels, newspapers, or empty feed bags (to lay beneath the kids as they are born)
- Paper towels, old towels, wash clothes, t-shirts (clean absorbent materials)
- Bulb syringe (to clean out nasal passages or throat in case kid cannot breathe)
- Flashlight (even if your barn is well lit, you may need the extra light)
- Scissors (to pop a bag of fluids, or to cut the umbilical cord – either of which are seldom needed, but you must be prepared)
- Lubricant (commercial brand lubricants or mild dish soap)

- Gloves (if you prefer, purchase long armed gloves from a farm supply store)
- Iodine 7%
- Small canister (shot glass sized for dipping umbilical cord in iodine)
- Unwaxed/unflavored dental floss
- Hair dryer (in very cold climates to dry the kid's coats)
- Heat lamps (in very cold climates, see page 185).
- Bottle and nipple
- Syringes and tube feeder for feeding weak kids
- Colostrum (preferably from your own farm, or a nearby farm, see page 186)

You might also consider a baby monitor or a barn camera (see page 35). I have never found a camera necessary, but an audio monitor has been helpful. The sound of a doe in labor is distinct due to her grunting, and as labor progresses some become more vocal. You will learn the voices of individual goats as time goes by, so if you have an audio monitor, you will know by sound and instinct who is "speaking," what they are doing, and if your monitor has enough volume, all of this can wake you up in the middle of the night – as an alarm, labor is on!

Instead of newspapers, or empty feed sacks, you may consider purchasing puppy pee pads to lay the newly birthed kids on. The pads are absorbent which helps with the process of cleaning and drying off.

Farm supply stores carry lubricant in a tube. I do not find the tubes convenient. When a veterinarian told me to use dish soap as a lubricant I was concerned. Would that hurt the doe? I decided to give it a shot. From the first time I used it, I loved it. It is easy to pour out copious amounts of dish soap, and it does not cause problems for the doe. If you do decide to use dish soap, select a type that is basic, designed for washing dishes and nothing else.

If your fingernails are long, clip them prior to the 140th day of your goat's gestation. If you have to go in and assist with birthing, you do not want to cause injury to the doe or kid. Also, remove all jewelry. Jewelry will be in the way, may get lost, and can cause injury to a doe or kid.

Nearing the Due Date

Have you heard the old saying, "A watched pot never boils"? As you wait for the onset of labor, you will understand why knowing the due date is so important.

Some does go into labor without outward signs. They push other goats out of their way at the hay or feed stand, eat with fervor, and then go into active labor just minutes later. While other does lay off feed for hours before they begin to have contractions.

First Stage Labor

There are three stages of labor. The first stage, when the kids enter the birth canal, normally lasts 12-24 hours. This stage can last for just a few hours, as well as several days. Much depends upon the doe, the breed, her body condition, and her age.

Remain calm through all of the stages of labor. Always trust your instincts. Like most animals, goats notice our emotions and reactions. If you are overly stressed, they will be too. When you are composed, it is easier to form a mental picture of what is going on in the birth canal, especially if you have to "go in" and turn or pull a kid that is presenting itself in a difficult position.

In this first stage of labor, the doe may begin nesting, digging through bedding, or she may merely paw the ground. She may stare or gaze off, or look wild-eyed. The doe may often yawn and stretch, acting as if she is bored with the entire process. She may frequently elevate the front end of her body by putting her hooves up on a fence or an object.

During this stage, she may look back at her sides, as if she is questioning what is happening with her body. The kids have shifted, and she feels and senses the movement.

Discharge from the vulva may appear. Do not base the actual start of labor on this. Only when the discharge thickens, and when it becomes more profuse, it is a signal of labor. Unless the discharge is bright red in color, which indicates the presence of blood, discharge is a normal indication of future or present birth. Thin discharge is common for weeks and months prior to birth.

The doe may also grind her teeth in this first stage. This is an indication of discomfort because the kids are moving into the birth canal.

When the udder tightens with milk, often called "bagged up," it can be a sign that kidding is near. Some does get their milk in early, days and weeks before they give birth, and milk does not come in for some until the day they give birth.

The softening of the ligaments is a physical sign of impending birth for most does. The ligaments are located just below the tailbone, angled in front of the tail head (which is where the tail is attached to the body), and are attached to the pelvis. Get in the practice of checking the ligaments of each pregnant doe on a daily basis. These ligaments are normally tight, but within 24 hours of the start of labor, you will notice a definite softening. When this occurs, the tail bone may become prominent, as if you can nearly wrap and touch your fingers together under the bone.

If your doe is panting, breathing fast or heavily, and she exhibits all or many of the above signs, she will soon enter the second stage of labor. At this time, the thin stream or discharge from the vulva normally becomes a much thicker stream of mucous.

Second Stage Labor

The second stage of labor includes contractions and pushing.

By this time, you should have the doe in a maternity pen. Have the necessary birthing supplies or kit within your reach.

When contractions begin the doe may lay down and push, stand back up for a few seconds or minutes, and lay back down again several times.

After contractions begin, the first thing you will see is a water bag. This is a tan-colored bag, which looks much like a bubble before it emerges. The bag may break, it usually does as birthing occurs, or may remain intact. If the bag breaks, the kid must be born because it will be unable to breathe once the umbilical cord separates from the placenta. Do not be alarmed if a second bag filled with darker fluids presents itself. After the water bag appears, the doe will lie down and begin pushing. She may be very vocal at this point.

The doe may stand back up and re-position herself between contractions.

Goats tend to birth in corners, close to walls, which often it makes it inconvenient for you to see what is happening with the birth. This is where a flashlight comes in handy.

After the water bag appears, and as the doe pushes, within a half hour you should begin to see the tip of a nose or hooves. The toes should be pointed downwards. If the toes are pointed upwards, or if one is pointed downwards and another is pointed upwards, this will not be a normal delivery and will likely require your assistance.

The most common birthing position is nose down with the front hooves positioned one on each side of the head, or right under the chin, with the head resting on the front legs. I call this the praying or diving position.

There is no reason to rush in and pull if you see that the hooves and nose are presenting in a normal birthing position. Some kids are born after the dam pushes multiple times, and some are born quickly with one push.

When the kid is born, the dam will begin cleaning the newborn's nose and mouth. This clears any mucous or birthing materials that may block the nostrils or mouth. I am not quick to pull a kid, or go inside to help with birthing, but I am quick with clearing the breathing passages. A clean, dry washcloth works well for this. I also run my clean or gloved fingers inside the mouth to clear any birthing materials. Normally, the amniotic membrane sac breaks during the birthing process, if it does not you will need to cut or break it so the kid can breathe

The dam should continue cleaning the kid at this time. If you want, you may assist by briskly rubbing the kid with a towel. In mild weather, I like to leave this job up to the dam. It helps her bond with the kid and it stimulates the youngster to get up and stand on its own. If you do rub the kid with a towel, do not be overly aggressive; yet, use some vigor to encourage blood flow.

With twin births, the dam will once again have strong contractions and the second kid is born soon after the first. If she strains and contracts for longer than 30 minutes, the kid may be in an abnormal birthing position (see page 175), and you may need to go in and assist.

During multiple births, consider moving the first born a few more feet away from the dam to ensure she does not lay down on it as she begins pushing. Always, if the dam is going to nurse a kid, return the kid to its place next to the dam as soon as possible. This keeps her attention on the kids.

After the doe gives birth, consider feeding her fresh hay, or even a small amount of grain. In colder weather, a bucket of warm water with a small amount of molasses mixed in will provide energy and will help her to regain strength, which promotes milk production.

Dip the kid's umbilical cord in iodine. The cord should be about 2" – 3" long. As long as the kid cannot step on the cord, there is no reason to cut it. Once the dipped cord dries, it will shorten in length. If bleeding occurs, tie the cord with dental floss to stop the bleeding. Wear gloves during the dipping process to prevent staining of your skin. The dipping process: pour 7% iodine into a small shot-glass sized cup or container. Hold the kid so the cord is centered over the container. Carefully dip the cord down into the container, up to the skin of the belly. This prevents disease and illness. The cord will dry out, shrivel, turn black, and will eventually fall off within several weeks.

Observe the kids to ensure they are strong enough to stand, and that they nurse. Check them regularly to ensure their bellies feel rounded – which is an indication that they are full with milk. Kids can go without nursing and survive for multiple hours before they go down and become weak. The sooner they have colostrum in their system, the elixir of a newborn goat's life, the better (see page 186).

Third Stage Labor

The doe has given birth, the kids are nursing, and she is finished, right? No, she will be expelling the afterbirth in the final stage. Normally, there is one afterbirth per kid, but occasionally there is a single afterbirth for twins.

Delivery of the afterbirth may take several hours. In some does, it occurs minutes after birth. Occasionally, the afterbirth will hang from a doe for several days. This is not a good scenario because the chance of infection increases. Never pull on the afterbirth. Always allow it to leave the body naturally. If the afterbirth has not expelled after several days, contact your veterinarian.

To prevent infection, administer Penicillin G to the doe for 2 – 3 days after you have assisted with birth, or when the afterbirth is slow to expel.

Your livestock guardian dog may eat the afterbirth. It is their instinct to rid the area of materials that draw predators.

The doe will have discharge for several weeks. It may look bloody in the beginning, and will stick to the underside of her tail, which is unsightly but will not cause health issues. The only time you need to be concerned is when the discharge has a foul odor, if it is pus-filled, green in color, or if it suddenly increases in volume after the first week. It may increase after a couple of days, which is normal as her body is eliminating any remaining materials and fluids.

The doe and the kids will bond during the next 3 to 7 days. While bonding, the kids will learn their mother's voice, and she will learn theirs. The kids will also copy their dam and may begin to eat hay at this young age. The chances of a kid not nursing are much slimmer when they enclosed in a pen with their dam.

The kids gain much strength during their first 24-48 hours of life. They quickly go from stumbling around to hopping about.

Abnormal or Difficult Birthing Positions

Most healthy goats give birth unassisted; however, eventually every goat owner experiences a difficult birth.

The following list includes numerous birthing positions.

- **Rear feet coming out first** - soles of feet are pointed up (considered a normal position, but not as common)
- **Butt first – breech birth** - kids can be delivered, sometimes not. If assistance is needed, push the kid back, maneuver one rear leg so it is coming first, and then maneuver the other leg forward
- **One front leg bent back** - kids can be delivered, but is very difficult on the kid and dam. Reach in and pull the bent leg forward so the kid is in **diving or praying position** (see Cornell University, page 247)
- **Head bent back** - cannot be delivered, and commonly stillborn. Reach in and push the kid back, get the head to face forward into **diving or praying position**

- **Multiple kids trying to come at once or are tangled** - cannot be delivered. Pulling may cause this situation to worsen. One or both kids may become lodged and may be born stillborn, with damage to the dam. Reach in and feel which parts belong to which kids, determine what positions they are in. Try to straighten them out. Determine which legs go with which kids before rearranging. Once you have a visualization of the positions, push one kid back to allow the first one to come out.
- **Tail first** - kid bent over, in the middle. This is not a good presentation, and is impossible for the doe to push out. Reach in and rearrange the kid. Time will be of the essence because in many of these presentations, the kid has swallowed birthing materials, the amniotic bag has busted, and the kid may have died.

Assisting with Birth

If you must assist with birth, which many farmers call "going in," roll up your sleeves, wash your hands, and if you prefer, put on long gloves. I prefer no gloves because I like being able to feel the position of a leg, tail, head, etc.

Take a deep breath and sit back for a moment. Do not scream, yell, cry, or do anything that may excite the dam.

Lubrication is very necessary at this point. You do not want to wait too long to go in, in case the doe is weak, or in situations where her contractions begin to weaken, or dilation decreases. In this occurs, call the veterinarian immediately.

To avoid injury to the doe or kid, make corrections to the kid's position between contractions, and pull during contractions. Always pull downwards, towards the doe's feet.

The birthing presentation may be normal, and the doe may simply need assistance with clearing the kid's head or shoulders. In this situation, grasp the area behind the kid's shoulder, which gives you something to hold onto, and gently pull forward while the doe is having a contraction. Never pull on a kid's limb.

Newborn Breathing Difficulty

Occasionally, especially following a difficult or breech birth, or when kids are born in multiples of more than two, one may come into the world only to have swallowed or inhaled birthing materials which prevents them from breathing.

Using a clean dry towel or cloth, clear the nostrils of mucous. Using your fingers, clear the front and back of the mouth of mucous or birthing materials

If the kid continues to struggle, swinging may help to clear birthing fluids from the breathing passages:

- Grasp the hind legs, directly above the kid's feet, with one hand (you may need to wipe their legs and feet off to remove slippery substances)
- Hold the head and neck with your other hand (this steadies and supports the neck area)
- By the feet, swing the kid back and forth several times with their head facing out
- Repeat these steps if the kid is still not breathing well

Swinging helps to clear the air passages, but if it does not work, or if you do not want to swing, use a bulb syringe to suck materials out of the throat and mouth.

The Next Steps

After the kids are born, the family will bond. Allow the doe to bond with the kids for at least three days, preferably a week if the maternity pen is large enough.

Observe the kids to ensure they are nursing and gaining strength.

If a dam rejects her kid, rub some of her colostrum or milk on the kid's mouth and rump. If she does not accept the kid, you will need to bottle feed (see page 188).

If a newborn is trying to nurse, and the udder is full, yet it appears colostrum is not flowing, a wax plug may be blocking the stream. This plug is natural protection against the entry of bacteria during pregnancy and is normally thin enough that it loosens when the kid nurses. Using a clean damp cloth, or a fragrance free baby wipe, wipe the ends of the teats and express the colostrum out. If the wax plug is unusually thick, you may need to massage the teats as well.

Our Birthing Stories and Learning Experiences

First Births on Our Farm

We purchased three bred does. We did not have an expected due date. The buck had lived in the pen with the does, and the actual mating was not been observed.

As the guesstimated due date arrived and passed, we waited and waited. We began to wonder if the does were pregnant. We had no idea what to expect in the latter stages of pregnancy, nor the hours before labor began.

The first doe finally gave birth to a single buck. She had a long labor, and no problems occurred. I assumed this length of labor was normal for goats.

When the second doe went into the first stage of labor I thought the birthing process would take just as long as the first one did. We were in need of supplies and we decided we had time to make a short shopping trip. Unfortunately, we should have stuck around. When we returned there was a large dead kid on the ground, still warm, with the doe looking at us as if she did not know that she had just given birth.

At this point, I was determined to not miss another birth.

I did notice the early signs of impending birth with the third doe. I made multiple trips to the barn throughout day, and even with my lack of experience, I knew it was going to be a long process, and that I had nothing to fear. Despite all of that, I was not going to miss a beat, so I slept on fresh straw in the barn that night. As the early morning hours approached, I finally dozed off. At daybreak, I woke to a barn cat snuggled at my head, and a doe completing her final pushes. I saw no water sac and there did not appear to be any discharge. I jumped up, in fear, and said, "No! You cannot push yet!" The kid emerged, a large doe, without any issues.

What lessons did I learn? No two births are alike. If you are able, do not leave a laboring doe alone, especially one that is a first-time mother.

Multiple Goats in Labor

As I accumulated experience, I found that is common for several does to go into labor at once. This is not a problem unless more than one doe has issues. Then, you may need a helping hand. Stay calm regardless.

One year I had four does in labor at once. It is easy to look at a doe across the aisle of the barn and feel that she having trouble giving birth, especially when you are squatted on the floor assisting another one. This is when you pick and choose which doe gets your full attention.

Unborn Kid Takes a Ride

One of the funniest stories from our farm is about a doe that normally needed help with each birth. She was a strong doe, with ligaments that never loosened prior to kidding, and with an udder that filled to the busting point, but not until the final hours before birthing. She produced large healthy kids, which was my decision point to allow her to breed year to year, despite the final pull that I always had to make.

That particular day I entered the barn to find her standing off by herself. She was vocal, in fact, she called out the moment she saw me.

I got close enough to observe what looked like remnants of a water sack hanging. I looked around for a dead or dropped kid. I found nothing, so I decided to put her in a maternity stall where she would be under my close observation.

Soon, she began pushing. One hoof emerged and she pushed again. Something seemed off, it was taking too long. I decided to go in and was relieved to feel the nose and head in the right position. I pulled the second hoof forward, placed my fingers around a shoulder blade and said, "Okay girl, when you push I will pull!" A beautiful doeling was born.

This was not the dam's first birth, and I could tell she was going to have another kid. Normally, she had her second kid with no difficulty. So, I stepped out of the pen for what was meant to be a moment, and did not shut the gate behind me. I assumed the doe would be too busy pushing to notice.

She began contractions and grunting. No hooves appeared. Not even one. Then I saw a tongue. Then a nose. No hoof.

I went in. I felt a head, shoulders, and no legs. The legs were not in a forward position. Before I was able to arrange the kid's legs, the doe pushed and just an ear came out. It was an unusual sight to see. She pushed again and the head came out.

I knew I needed to move quickly, the kid was not coming out easily, and I still had not found the legs, so I cleared the kid's nose and mouth and it took a deep gulp of air.

At that very moment, the doe got up and shot out of the pen. She ran across the barn, right to a pile of dirt.

As she ran, I could see the kid looking around, as if it were going for a ride. I heard my own laughter as I said, "You, are one crazy goat chick!"

Long story short, I finally got the doe to lie back down and I found the kid's folded legs and pulled them forward. During the does contractions and pushes I pulled. A beautiful healthy doeling was born.

What did I learn? Never leave a gate open, even when a doe is in deep labor. Never depend upon loose ligaments, or a full udder, as an imminent sign of impending labor. Never assume that any labor is going to be the same.

Angry Kid Reunited with Mother

The doe had given birth to a large and healthy doeling. I missed the birth entirely. The newborn kid was clean and dry, so I knew she was at least several hours old. The kid appeared to have strength, but I was concerned when I picked her up and felt her belly. It seemed soft, flat, as if she had not nursed.

I decided that time was on our side. I would monitor the kid to see if her belly filled out, and I hoped to catch her nursing. Much to my dismay neither happened, and when I looked down at the dam and I saw no trace of an udder. She was a dairy breed, so this made it even more unusual.

With sadness in my heart, I was exhausted from a busy week of birthing, I lifted the kid out of the pen and took her to the house where I would heat colostrum and begin bottle feeding.

The kid put up a fight. She would not nurse. She seemed angry. She had spunk. This continued through the night.

When the light of morning made its appearance, the kid was still fighting the nipple. She absolutely refused to nurse. I was not happy with my lack of success, and I was amazed that she was so strong.

I made my way to the barn to feed and milk. The last doe that I put on the milk stand was the mother to the kid that was still in our house.

Much to my surprise, I found a small and tight udder that sat further forward than most. She had colostrum, and I instantly knew why that kid had been mad! She had nursed, and that was why she was so strong.

I ran to the house, retrieved the kid, and she headed straight to her dam's udder as soon as I set her down in the pen.

What did I learn? I should have observed more closely to see that the kid was healthy and nursing. Instead of looking for an udder that would have been visible from behind, I should have reached down and felt with my hands. We do not want to breed does with irregular udders, because we would then produce kids with irregular udders, but these are indeed strong and healthy animals. That should have been my first clue with this particular kid, she was already strong and healthy!

Our Experience with Birthing and C-Sections
I personally experienced all of these described birthing positions. The first three positions listed in this chapter, along with the praying position (on page175), are the most common.

I have taken a doe to the veterinarian for an emergency C-section. She and I laid in the bed of a pick-up truck in freezing weather for the 30-minute trip. She and her kids survived but they were weak for many days.

When C-sections occur, ask the veterinarian if the doe should be bred again. The veterinarian will tell you if the problem is related to the pelvis or bone structure, which could result in future birthing issues.

Chapter 12
Raising Kids

While the buck is half of the herd, kids are a large segment of the other half of a production or income-bearing herd. Solid kid care is an investment in the future of a successful farm.

Kids serve a myriad of purposes – income, replacement animals, pets, and offspring to increase the herd size. Kids, in my opinion, can also bring great joy and entertainment. Like a young child, they seldom stand still, and they break us out in smiles!

Newborn Care (First 24 hours)
The naval is a direct route for pathogens to enter the newborn's system. Dip the naval as soon as possible following birth (see page 173).

Keep newborn kids in a clean, dry, and draft free environment. Straw bedding is preferred. It does not cling, and it provides warmth when layered thick.

Heat lamps are helpful when raising newborns in cold climates; however, they are a fire danger. Never leave heat lamps unattended while on, and never place them within the reach of animals.

The first 24 hours of a kid's life are critical. They are born with little or no immunity to pathogens and bacteria in the environment, nor can they withstand adverse temperatures. They depend upon their dam's colostrum for some immunity and for their own internal ability to produce and preserve body heat.

Goats are not born as ruminants. They begin life with only one of four stomachs, the abomasum. They rely entirely on milk for nutrition and energy for the first 7-14 days. When the kid begins to eat dry feeds, the rumen and reticulum launch into development, and as the microbes slowly increase, the ruminant system gradually grows. When a very young goat eats solid feeds, it goes into the rumen, which helps it grow; however, when they nurse the suckling reflex signals the body to bypass the developing rumen. The esophageal groove closes, which sends the milk directly to the abomasum, where it is digested. The remnants then move on to the intestines. Because the ruminant system is still developing, do not wean kids from milk at a very young age.

Feeding the Newborn - Colostrum
There is nothing more important to a newborn's health and mortality than colostrum, which is a thick, sticky, yellowish substance that a dam produces at birth.

Liquid gold is the name I have chosen for colostrum. It is the single most important food to store on a farm for future kids. I keep a minimum of 16 to 32 ounces of colostrum from my own farm in the freezer, year round, just in case I have a newborn kid to bottle feed. Note, I date the bottles and use the oldest colostrum first, with a use by date of a year or less. To obtain colostrum for the freezer, heavy milkers produce enough colostrum for their kids, with extra to milk out. In addition, if a doe loses her kid at birth, milk out her colostrum, and heat treat it as desired (see page 218). Then, bottle the colostrum and freeze.

To avoid colostrum deprivation, which leads to a weak kid or one with poor immunity, a newborn must receive at least 10% of its body weight in colostrum within the first 6 to 12 hours of birth. My rule of thumb is to ensure a newborn receives some colostrum within the first 2 to 3 hours of life. The absorption rate of the protective qualities of colostrum drop considerably after the first 6 to 12 hours of birth.

For immunity purposes, if a newborn kid cannot nurse its mother for any reason, the best solution is to milk the mother and feed the kid via a bottle, or tube feed (see page 188). If you cannot milk the dam, feed colostrum from another goat from your farm. It will contain anti-bodies that are unique to your farm. If these are not options, obtain goat colostrum from a nearby farm, and heat treat (see page 218) prior to feeding.

Goat colostrum is available in powdered form. It does not do not provide any life-saving protection from disease; however, it does provide initial nutrition and energy. If powdered colostrum is your only option, consider mixing it with goat milk from your farm to help the newborn form antibodies.

If a newborn kid is chilled, unable to nurse, never attempt a forced feeding. To ensure the organs in the body are warm enough to function properly, which will allow digestion and absorption of the colostrum, the kid's internal temperature must fall within the correct range. I have outlined the procedure that we follow for reviving chilled newborn kids on page 197.

Never heat colostrum above 135 degrees Fahrenheit. Higher temperature will cause the colostrum to solidify and the heat may destroy its protective anti-bodies. When thawing a bottle of colostrum, leave the cap on and place it in a container of warm water. It will thaw rapidly.

A dam produces colostrum for 2-3 days following birth.

Urination and Bowel Movements

A newborn kid begins urinating and having bowel movements soon after it is born. Once it starts nursing, you will want to make sure the digestive system is working properly by paying attention to the color of the bowel movements.

Colostrum acts as a laxative, in addition to the nutrients and antibodies it carries, it helps the newborn's body excrete the first feces, called meconium. For the first day or two following birth, the bowel movements will be dark in color and very sticky. The feces will then change to yellow, and will eventually turn brown in color, especially after the kid eats hay and solid feeds on a regular basis.

The dam keeps her kids clean by licking, which also promotes the natural flow of feces, but there are times when the yellow bowel movements harden and stick to the kids, which may lead to blockage. Gently pull the stuck feces off.

Why Bottle Feed?

Bottle feeding is the sustainer of life when a very young goat cannot nurse from its dam. If you breed goats, keep bottle feeding supplies on hand: bottles, nipples, frozen colostrum and goat milk (milk replacer if you have no fresh or frozen milk).

Many dairy goat farmers remove their kids from their mothers at birth. They do this to protect the udder from damage, and they feed heat treated colostrum to protect the newborn from disease. Many cheese makers utilize all of the milk that is produced, and they feed milk replacer after kids are several days old.

Necessity is the most common reason goat farmers bottle feed - the newborn kid was initially too weak to nurse, the dam became ill and unable to nurse, or she rejected her kids entirely.

Some first-time mothers will not know what to do with their newborn. She will give birth and look back at the kid as if it came from another planet. In these cases, do not step in and clean the kid completely off. This leaves the scent of the doe on the newborn, which helps the mother recognize that the kid belongs to her. If her natural mothering instincts do not kick in within the first hour, you may need to calm the mother, or secure her, and gently urge the kid to nurse from her. Introduce the teat into the kid's mouth and help by squeezing out a drop or two of colostrum. Once the kid begins nursing, the battle is normally won – the dam and newborn will bond. Some mothers will reject, and even beat up their kids as soon as they are born, but this is not common. In this type of situation, you will be forced to remove the kids and bottle feed.

Lastly, there are positive results from bottle feeding kids. They generally become gentle, loving, as if they assume that you are their parent. Once a bottle kid, always a bottle kid, in other words, they will follow you around much like a young child or loyal pet, throughout their entire lives.

Bottle Feeding - Precautions

Never over feed a kid goat. In newborn and young goats, what we think we are doing out of love (feeding too much and too often), can kill. Stick to a feeding schedule. Unlike a human child, when a young goat cries, it normally does not indicate hunger. Do not stick a bottle in its mouth unless it is feeding time!

If you do not have access to goat milk, select a well-balanced milk replacer. The milk replacer label must state that it is suitable for kid goats, and the replacer must contain copper (a very necessary mineral for goats). When mixing the milk replacer, carefully follow the package instructions. To avoid illness, it is much better to add slightly less replacer to the warm water than it is to add too much. Finally, be consistent with your measurements each feeding. Remember, gradual changes in diet allow the system to adjust.

Bottle Feeding – How Much and When

When I stepped into the world of bottle feeding, I fed until the kid's belly felt full, or until they began playing with the nipple, which indicated a lack of interest or fullness. I fed up to 20 ounces per bottle, 3 times a day. It seemed that the bottle fed kids became ill more often than the dam fed kids. I am convinced this was caused by overfeeding and it lead to kids that had little, or no, interest in hay or solid feeds.

To get an idea of how much to feed a kid goat, watch one nursing on their mother. The dam allows the kid to frequently nurse, but not for extended times. She allows the kid to drink a little (almost a "slurp" as we lovingly call it), then she makes the kid stop. Hence, when we bottle feed large quantities of milk, we are allowing the kid to drink more than he would if he were "on" his mother.

Bottle Feeding Schedule

Day One   Always feed colostrum! Up to 6 ounces per feeding, every 4 hours

Day Two Colostrum - up to 8 ounces per feeding, 4 times a day (can begin mixing small amounts of milk or replacer with colostrum after 36 hours).

Day Three Colostrum mixed with goat milk or milk replacer. 10 ounces per feeding, 4 times a day (gradually lower the amount of colostrum in the mix).

Day Four Colostrum mixed with goat milk or milk replacer. 10-12 ounces per feeding, 4 times a day.

Next Two Weeks

Goat milk or milk replacer. 10-12 ounces per feeding, 4 times a day.

Up to 2 months old

Goat milk or milk replacer. 10-12 ounces per feeding, 3 times a day.

Up to 2 1/2 months old

Goat milk or milk replacer. 10-12 ounces per feeding, 2 times a day.

At 2 1/2 months old begin weaning (see page 194). Lower the amount of milk per feeding by about an ounce per day (or two ounces if the kid eats hay regularly). This encourages the kid to eat solid feeds.

Note: The amounts of milk or replacer in this schedule is based on medium to heavy breeds of goats, feed less if you are feeding a small breed.

Always provide clean water. To prevent drowning, do not place deep buckets or troughs on the ground in small confined areas, such as in the maternity pen.

Bottles, Nipples, and Nursing Tips

Kids generally take to the bottle with ease, unless they are weak or ill, or already nursing from an udder.

Sit down with the kid in your lap. Have them lay in a position with their head held up. They may also nurse standing up. Picture how they naturally get under an udder and reach up to the teat. This is the safest nursing position, which helps to ensure the milk does not enter the lungs, and it helps to prevent the swallowing of air. Do not be surprised if the kid stands and then kneels while nursing, both of which are safe positions.

If the kid plays with the nipple, or tosses their head from side to side in an act of refusal, try applying a drop or two of white corn syrup to the tip of the nipple. The sweet taste alone may entice the kid to begin nursing. Try inserting the nipple in the kid's mouth and gently hold the sides of their mouth closed while rubbing their throat. This nearly takes more than two hands, but practice makes perfect!

Squeeze a drop or two of the milk into the kid's mouth. Once they realize there is something tasty in the nipple they usually start nursing. Remember, never force feed a weak or newborn kid.

Can kids be pan fed? I had one that insisted on pan feeding. He did eventually take to the bottle. I do not recommend pan feeding because young kids do not digest milk in their developing rumen, and the physical act of nursing detours the milk directly to the abomasum where it is fully digested.

I prefer nipples that screw on tight to plastic (water or soda) bottles. I only use the plastic bottles once. Plastic tends to etch, which makes them difficult to clean. The nipples that I use are closest to the feel of a real teat, and they have a valve that controls the flow of the milk. Ensure you have enough nipple replacements on hand. Some farm stores discount when a dozen are purchased.

A lamb or milk bar is another bottle feeding option. It consists of a vat that holds the milk, with tubing, valves, and nipples. This method of feeding allows multiple kids to nurse when they want or need to, which frees your hands up. Typically, kids take sips from the milk bar feeders in short intervals, as they would when nursing from a dam.

Introducing the New Family to the Herd

When the kids are between 3-7 days of age, they have gained strength and have bonded with their mother. At this point, you may introduce the family to other goats. This decision should be based on how well both the dam and the kids are doing. Are they eating well? Are the kids walking without issues?

Adult goats have a tendency to spat after they have been separated for several days. This adjustment period will pass, but initially keep an eye on the kids.

At this point, you should have already checked the fence lines to ensure there are no escape points.

Feeding Solids

Solid feeds consist of any type of feed that is not a liquid.

I prefer offering good quality legume hay, especially alfalfa, or an alfalfa blend, in the first weeks of life. I do not offer concentrates or grains until the young kid is at least three weeks old and after it is eating hay. Then, I only offer grain in very small quantities, less than ¼ cup per kid (medium to large breed), if at all. Grains and concentrates contain high percentages of starch. An immature rumen cannot digest an influx of starches. Hay, however, helps the rumen develop.

Kids will often pick at hay, and not ingest much of it in their first few weeks of life, and they normally do this at their mother's side. Kids that are on a bottle, separated from their mothers, tend to prefer a bottle with little or no interest in solid feeds. To entice a bottle kid to eat hay, bring in an older kid during feeding to set the example.

Any time you combine kids in a pen, especially in a common feeding area, ensure the smallest of kids are able to eat – not starved or bullied out. If several of the kids are larger, or stronger, consider moving them to a separate area where the kids are uniform in size.

Consider a creep feeder when kids are in a joint area with larger goats. A creep feeder is a trough or hay rack enclosed in an area that only the kids are able to enter through a small opening.

Kids like to bounce in and out of feeders. This contaminates both the hay and the trough, and can be troublesome for parasite control during damp seasons. Set creep feeders up in dry and easy to clean areas. Do not set out so much hay that it becomes contaminated day after day. Fresh and clean is what you want to achieve.

Weaning

We wean our bottle kids between 2 1/2 and 3 months old. We allow the dams to wean their own kids when they desire. Use caution, though. Young kids are fertile early in life. For this reason, we separate bucks and does when they are 2 months old.

When kids are being weaned they become vocal. They may also take on a thinned out appearance for a day or so. Ensure they have fresh clean water and good quality hay. Consider introducing a small amount of grain to supplement through the weaning process.

Disbudding (Dehorning)

Disbudding is the removal of horn buds of young goats (see page 149). If you plan to disbud, you will want to do so before the horn buds have grown considerably in size.

Castration

Castration, also called neutering, is the removal of the testicles. Castration prevents the goat from breeding, and in some cases, it helps to control aggressive male behavior as the animal matures. A castrated male goat is called a wether.

Castration methods:

Banding (Elastrator) – A heavy rubber castrator band or ring is applied. It cuts off the blood supply to the testicles. It is a bloodless method but does cause some pain. There is no open wound, which leaves little room for infection. This is the most common method of goat castration.

Burdizzo – A metal instrument is used to sever or crush the blood vessels and cords that lead to the testicles. There is no open wound, which leaves little room for infection.

Knife – This is a traditional method, not used as often on small farm. This method leaves an open wound, which can lead to infection and fly infestation.

Do not castrate before the buckling is 30 days old. This allows the urethra to grow, which helps to prevent Urinary Calculi (see page 117).

When following any of these methods, I recommend the involvement of a veterinarian. He will explain how to band, and will perform any of the castration methods for you. The veterinarian will also provide prescription pain medications.

Before castration or disbudding, ensure the animal has been vaccinated for Tetanus.

Vaccinations

Vaccinations prevent disease, which is extremely important as young goats are building their immune system (see Enterotoxemia, page 86, and Tetanus, page 115).

***

Special Care

Weak Newborn

Most kids will rise up on their wobbly legs and nurse within the first hour of life, and some will pop up within minutes of birth and nurse with miraculous gusto – as if they were a week old already. Then, unfortunately, there are weak newborns that are unable to stand or nurse. Other than a chilled newborn kid that is in grave danger of death (see Reviving a Chilled Newborn, page 197), there are several points to consider when working with a weak newborn:

- Help the kid stand up. Raise them up by their belly and straighten their legs out, which encourages them to use their legs and to gain strength.

- A weak kid may be lacking in Selenium, and one dose may work like a miracle (see White Muscle Disease, page 119).
- Watch the kid for symptoms of disease. A newborn, however, is not normally born with disease. Illness sets in when a newborn is deprived of colostrum. Disease occurs later in life after exposure.
- Slings, either handmade or purchased, work wonders for kids with weak legs. Place the kid in the sling, low enough for the kid to put all four hooves on the ground and stand, but not so low that the kid sits down without making an effort to stand. Monitor the kid while they are in the sling to avoid dangling or strangling. Provide this physical therapy several times a day to strengthen legs.

Reviving a Chilled Newborn

You walk into the barn and find a limp newborn that has entered life in a difficult way. Perhaps the kid was weak to begin with, but is now chilled, wet, and unable to stand. The newborn has not had colostrum and is nearing death.

The newborn is lethargic and limp, with slow respiration, and may be gasping for air. Its head may be drawn sharply to the side, and it may have a weak heartbeat. In these situations, mere minutes remain in the newborns life.

Wrap the kid in a towel or small blanket, talk to it, gently rub its sides, and whisk it off to a climate controlled building that has a sink with warm running water.

Remember, the normal goat temperature range is 102-104 degrees Fahrenheit. The use of a thermometer is always the most accurate method for temperature reading, however, time is not on your side in this situation. Insert your finger into the newborn's mouth. If the kid's mouth is chilled, then their entire body is chilled as well.

A newborn goat kid will not retain its own body heat until it has had sufficient nutrition (see Colostrum, page 186), however, a very chilled goat kid cannot, and will not, nurse. Do not attempt to drip liquids down their throat. If they are chilled, to the point they cannot nurse, the liquids will more than likely reach their lungs and drowning can occur, or pneumonia will later set in. Internal organs, including the abomasum – where initial nutrition will be absorbed, will begin shutting down as the body temperature plummets.

Place the kid in a sink of warm water (close to 102°F-104°F degrees). Again, time is not on your side, so simply make sure the water is comfortably warm. Keep the newborn's nose and mouth out of the water. A divided kitchen sink works best for this, as it gives you a smaller area to work with. Maintain the water as close to temperature as possible. If the kid is very chilled the water may cool down quickly, replace it as often as needed. If possible, keep the kid immersed while adding additional water.

I keep a bottle of maple, corn syrup. or molasses near the sink during kidding season. As I hold the kid in the water, I pour a couple of teaspoons of syrup into a small bowl or saucer. I dip my finger into the syrup and rub a small amount into the insides of the kid's cheeks. Repeat several times.

As the kid's body temperature nears a safe level, they will begin to jerk or shiver. Until the kid begins to do either, their body temperature is probably not climbing. You will notice that the inside of their mouth warms as they begin to shiver.

Once the kid is more alert (temperature nearer to normal) I pull them from the warm water, quickly wrap them in a dry towel, and begin drying their coat with a hand held hair dryer. Caution, light massage is a good thing, but never become overly rough with towel rubbing. Kids can only take so much when they are already exhausted. When drying the coat, to avoid burning the skin, do not hold the dryer too close, or in one position. Use sweeping movements as you dry. Gently massage the kid with one hand, while holding the dryer with the other.

Once the kid is dry, after following these steps, the body temperature should have increased. They will be more alert, but may still be showing signs of impending death. Their blood glucose level is very low at this point.

Begin administering liquids through injection. Using a 20 gauge needle, 1/2", inject Dextrose (use a 5% solution, and/or the Dextrose solution manufactured specifically for injections), SC (under their skin), in the area between the shoulder blades. Pinch the skin to form a "tent" and inject the Dextrose just under the skin. The kid can survive on the SC liquid for several hours.

Inject 3 ML Dextrose, repeat, until a "hump" is raised between the shoulder blades. Within 15-20 minutes, you will notice the hump going down. This means the body is absorbing the liquid, which is a very good thing. Until the kid has fully revived, repeat these steps as new fluids are needed. The kid's body will absorb all of the hydration it needs through the injected liquids.

Consider keeping a bag of Lactated Ringers on hand (a bag of saline IV solution purchased from the veterinarian). In extreme cases, rotate the injections between Dextrose and Lactated Ringers.

Keep the kid wrapped in a dry towel, and make sure they are in a warm location free of drafts. Revival of a chilled kid is a gradual process. Do not expect miraculous results in a matter of minutes.

Again, do not try to force liquids down the kid. They will develop pneumonia, and they cannot nurse (or swallow properly) until their body functions resume to a normal level. Once the kid is hydrated and their body temperature is normal they generally want to nurse.

Remember to talk to the kid as you work on them. They respond to voice. Goats give up easily when they do not feel well. Your voice may provide just be enough reassurance to help them cross the bridge to a healthy long life.

As you follow these steps through the revival process, you will feel the heartbeat (just by touch), respiration (by watching the nose and the chest), and you will notice how the kid looks in the eyes. Goats have an angry and dull look to their eyes when they are not well.

If you have warm, not hot, caffeinated coffee on hand, you may also rub a small amount, a drop of two, between the inside cheeks, alternating with the drop or two of syrup or molasses. The caffeine and sugars act as a stimulant and may help increase the heart rate and blood flow, and will provide a small boost of energy.

Tube Feeding

Tube feeding is necessary when a young goat cannot nurse. Remember, their body temperature must first be brought up to normal, or as close as possible, for the milk or colostrum to absorb – body organs may not be fully functional when the kid is extremely chilled.

I personally prefer to leave tube feeding as a last option because there are two major dangers – aspiration (liquids enter the lungs), and injury to internal organs and linings.

Tube Feeding Supplies
- weak kid feeding syringe (60 cc/2 ounces)
- sterilized tubing
- warm water (approximately one cup)
- warmed liquids that you will be feeding
- two extra syringes (6cc) for measuring the amount of liquids you will dispense
- small container of warm water to dip the tube into

Tube Feeding Instructions
1. Measure the distance from the nose of the kid to the center of the ear. Measure from the ear down to the last rib (the chest floor). Add the two measurements and mark the tube (with a marker or piece of tape) the summed distance, from the tip up. The tip will be inserted first, and the mark is the maximum distance you will be inserting.

2. Place the tip end of the sterilized tube in warm water to soften.

3. Lay the kid across your lap or on a solid surface (table, counter top). Use your left arm to support the kid, or support his neck and head. These procedures are for right handed people; reverse the direction if you are left handed.

4. Hold the kid's head firmly just under the base of the ears. Retain your grasp and slightly lift the kid upwards and tilt the head back slightly (picture a hanging position).

5. Dip the tip of the tube in warm water to soften and help glide it into the esophagus. To eliminate the chance of contaminating the lungs, never use oil or any other substance to coat the tip.

6. With your right hand, guide the tube, without the syringe on the other end, slowly down the center of the mouth, over the tongue and down the throat.

7. If the kid is coherent, it may make swallowing movements, and if the kid cries, you most likely have inserted the tube into the correct passage – the esophagus. If the tube is inserted incorrectly into the trachea, he will most not be able to cry. While inserting the tube, if the kid can initially cry and suddenly cannot, if the tube stops short, or if the kid starts to choke, stop and slowly withdraw the tube and start over. To avoid tissue damage, **insert the tube only to the mark that you made.**

8. Once the feeding tube is fully inserted, put the open end of the tube, the end that you will insert the syringe into, in the container of water. If you see bubbles, slowly remove the tube from the kid and start over. Bubbles indicate that you are in the trachea. If no bubbles are seen, proceed to the next step.

9. **For older goats only,** smell the open end of the tube for rumen odor. A newborn does not have a developed rumen so you will skip this step.

10. Place the end of the tube at your ear and listen for breathing sounds. Breathing sounds from inside the lungs sound like crackling noises. If no breathing sounds are heard, fill the 6 cc syringe with 5cc of warm water.

11. Inject the 5cc of water into the tube with the 6cc syringe. Make sure it flows down the tube. If it does not flow, gently pull the feeding tube up by a few inches and try again.

12. Pull 2 to 4 ounces of the feeding liquid into the syringe and attach the syringe to the end of the tube. Gravity will deliver the fluids to the stomach when you hold the syringe up above the kid. Do not use the plunger, the middle part of the syringe. The fluids suction down the tube easily if you remove the plunger during this step.

13. Add another 10 cc of water to the syringe. This step prevents the kid from aspirating while the tube is being removed.

14. Remove the syringe from the end of the tube. Remove the tube slowly while holding your finger over the open end. Do not remove your finger, which prevents fluids from entering the lungs.

15. Clean and sterilize all supplies.

I recommend on-site tube feeding training before attempting this on our own. Visit a farmer that is well-versed in tube feeding, ask your veterinarian for assistance or a demonstration. At the very least, watch online videos of the process. Many higher education facilities have excellent pictorials of tube feeding on their websites as well.

To prevent damage to delicate lung tissues, never blow air into the tube as a test, nor should you suck air from the lungs into the tube. Some goat owners do this to ensure the tube is not in the lungs, which is a personal preference that I do not follow.

When inserting the tube, any time you reach resistance, if the tube does not slide in smoothly, remove the tube and start over. Resistance, coughing or signs of choking, normally indicates that you are in the lungs.

To avoid tissue damage, never remove the feeding tube quickly.

You can tube several types of liquids – electrolytes, 50% dextrose, colostrum, and goat milk. Remember, colostrum is the first and most important feeding of a newborn goat's life.

Never tube feed more than a young goat would normally ingest. Never tube more than 20-30 cc's liquid at a time, and if necessary, repeat tube feeding every 2-4 hours.

Chapter 13
Milking

Have you heard the saying a dairy goat is a poor man's cow? I say a man is poor if he has never had the wonderful experience of owning a dairy goat.

Picture yourself sitting on the milk stand with a goat at your side. She is munching away on grain, nearly humming as she goes, while happily dropping milk into your bucket. Gently lean against her side and talk to her. She will adore it. Cup her ears in your hands and give her a thank you with a nose to nose touch before releasing her back to the pasture. She is secure, and she gives all that she can through her milk.

The Udder and Teats

A goat has one udder, which has two compartments, each with separate mammary glands that produce milk.

A goat has two teats, one for each side of the udder. Teat size varies amongst breeds and by genetics. Some teats are very small, making it difficult to milk by hand, and some are large and bulbous which can create problems with machine milking.

It is advisable to not purchase dairy goats that carry a third or false teat, which can make milking difficult, or may promote mastitis, however, some meat goats, particularly the Boer goats, often have a third teat.

At the end of each teat is an orifice. The orifice opens when the goat is nursed or milked. Small orifices result in smaller streams of milk, which creates no additional problems other than milking will be slower for this goat.

The ligaments that hold the udder up close to the body are called attachments. The rear attachments hold the udder up, and the fore attachments hold the udder up and raise it slightly forward. You will notice that the teats on a tight udder normally angle forward.

Udder attachments are important in the show ring because they make or break the look of an udder. Attachments are also important to the long-term milking ability of a goat.

All about Goat Milk

About 65% of the world's population consumes goat milk. Why is that? Goats are manageable; they require less pasture and feed than cows. One dairy goat in full production can provide enough milk for a small family.

Goat milk does not need to be homogenized because it contains smaller fat globules, as compared to un-homogenized cow milk with cream that readily separates and rises to the top.

People sensitive to cow milk often find goat milk to be an alternative. The small molecular structure of goat milk makes it easier and quicker to digest with less, or no, symptoms associated with lactose intolerance.

Cow and goat milk is similar in fat content. On average, goat milk contains between 3% and 4% fat. Goat breed influences the fat content of milk. For example, the Nubian breed will produce high fat milk, to the point where the cream separates from the milk, while milk from the Alpine and Saanan breeds contain less fat with little to no separation. Remember, genetics and feed are also factors in milk fat.

Compared to cow's milk, goat milk contains more nutrients: 350% more niacin, 13% more calcium, 47% more vitamin A, 25% more vitamin B6, and 134% more potassium. Additionally, goat milk has more copper, manganese, chloride and selenium.

Goat milk may help with allergies, heart disease, diabetes, arthritis, and the list seems to be growing. Always consult with your physician, these facts are not to replace or alter professional treatments.

The taste of goat milk is affected by what the goat eats, by sanitation, and by her environment. A goat housed near or with a buck in rut may produce milk that is slightly "off" in taste. A goat that eats wild garlic, or a bitter non-poisonous herb or weed, may produce milk that tastes like the greenery that she eats. Keep this in mind when pasturing goats. Different hay, depending upon the type and quality, will also greatly affect the taste of milk. Legume hay, especially rich alfalfa, generally produces sweet milk that does not taste tainted.

Milking Frequency and Production

Dairy goats lactate for approximately 10 months, producing an average of 3 quarts a day. The length of time in milk, and the volume, varies based on age, feed, general health, genetics, breed of goat, and milking frequency.

When a doe freshens (gives birth) she produces colostrum for 2 to 3 days. Freeze the colostrum that you collect for future kids. Do not mix colostrum with goat milk if you use it for cheese, soap, or for your family to drink. The taste and consistency of colostrum differs greatly from raw milk, and it rapidly solidifies when exposed to high heat.

Most does are not in full milk production on the day she gives birth. Production gradually increases over a course of several months. After the first two months in milk, production will slowly decrease. The natural weaning period for a goat is 2 ½ to 3 months, with some that continue much longer. In nature, as the nursing kid grows, the volume the dam produces increases, and then towards weaning time the mammary glands produce less.

Most dairy farmers milk twice a day at a scheduled time. The modern farmer, one that works both off and on the farm, may find it easier to milk once a day. Either way, the key is to follow a consistent milking schedule.

When milking twice a day, at 12-hour intervals, the amount of milk collected will generally be the same at each feeding. If you milk twice a day, for example, at 9 o'clock and 4 o'clock, the volume of milk in the morning will be much higher than that collected in the evening.

When a dairy goat is milked much later than her scheduled time, or if milking is skipped altogether, hormones kick in which signal her body to slow down and stop producing. For this reason, do not deviate from a milking schedule more than 45 minutes to an hour. It is very difficult to increase a doe's milk production once her body makes the choice to stop producing.

Plan the milking schedule before your dairy doe gives birth and begin it the day she gives birth. The schedule that you follow will affect her milk production throughout the entire season she is in milk, and throughout her life. A dairy goat that is not milked regularly may never produce well. Also, a dairy goat that is not bred when she first matures will not produce as much milk throughout her lifetime. For this reason, if your goal is milk production, plan to breed your does as soon as they reach maturity and growth requirements.

Machine or Hand Milking

Machine milking is an option for a larger herd, or for one that produces milk for sales or cheese making. Milking by machine is also an option for someone who suffers from a medical condition that affects the strength of the fingers, hands, or wrists.

Hand milking is enjoyable to many goat owners. Often, it is quiet time that fits into a peaceful farm lifestyle. Whether you hand or machine milk, milking is an opportunity to handle your does on a daily basis. Milking, in my estimation, is bonding time. A goat that is on a collar, regularly lead to a milk room or area, and onto a milk stand, will cooperate with you, and will want your company.

Things to Know Before You Begin Milking

Before we talk about milking procedures, know that cleanliness and sanitation is at the top of the list of importance. Milk is delicate. Any amount of contamination may spoil the entire batch, resulting in tainted milk that can lead to illness. Contamination results from dirty teats, a dusty or dirty milking parlor or building, insects, or containers and milking equipment that is not sanitary.

Goats must also be calm during the milking process. I do not know how they do it, but once they start shifting around, or if they are startled during milking, the milk simply stops flowing. For this reason, provide a milking area with a stanchion, away from the rest of the herd.

Milking Supplies

Whether you hand or machine milk, you will need basic supplies:

- Clean area in which to milk. A milk room or parlor is the preferred place, but any dust free area sectioned off from the herd is suitable.
- Stanchion or milking stand with feed bucket or pan. Stanchions can be purchased pre-made, or made from simple instructions (found on the Internet), from wood or entirely from metal.
- Seamless stainless steel bowl or milking bucket. This should be shallow enough fit under a goat with a lower hanging or full udder, but tall enough to prevent splashing out. I prefer milking containers without lids. The size and shape of an udder, and teats, make milking into the same container for each goat difficult unless the bucket or container has a wide opening.
- Paper towels to wipe the udder and teats. Use the type of paper towels that are thick and strong, similar to soft fabric.
- Bucket of udder wash. Use a pre-made udder wash, sold by the gallon in farm stores. Or, use hand-made udder wash (see page 222).
- Unscented baby wipes. These replace the paper towels and udder wash. To prevent the udder and teat skin from chafing, use wipes that do not contain alcohol. I appreciate baby wipes in cold weather because it eliminates the need for a bucket of warm water, and the less water that is used on the teats and udder, the less chance of cracked dry skin on the udder or your hands.

- Brush for the goat's coat.
- Udder balm. Apply udder balm to chafed or irritated skin after the milking process. Always use teat dip or spray to help the orifices close before applying the balm.
- Teat spray or dip. These products sanitize the teats, and help the orifices to close after milking, which prevents bacteria from entering the teats.
- Scale. A scale is not a necessary item. Some farmers weigh each animal's milk and keep a record of productivity.
- Milk record sheet. This is not a necessary item, but you may want to keep a record if you weigh the milk. Record keeping can be helpful if you decide to sell the goat, and it may help you to decide which does to keep or breed for production.
- Stainless steel milk tote. Milk totes are available at farm stores in a wide range of sizes.
- Strip cup. A strip cup should be deep enough to ensure the milk does not splash out when the stream hits the cup. A stainless steel strip cup is easy to clean and light weight. A light or cream colored cup is not advisable because the it will make examination of the milk difficult.
- Glass jars for milk storage. Sanitized glass jars with tight fitting lids are extremely useful on a dairy farm.
- Plastic bottles or zip lock gallon bags for freezing milk.
- Stainless steel strainer
- Milk filters. Filters are absolutely necessary. They filter out tiny particles of most anything – hair, dirt, and visible specks of dust.
- Sanitizing solution for milking equipment

- Large container for ice bath – a heavy-duty plastic tote or canning pot works well. An ice bath is not necessary, but helps with the safe handling of raw milk.
- Mastitis test kit

Pre-Milking Instructions

1. Secure the goat in the stanchion or milking stand. She will be more than happy to put her head through the gate if you have placed a small amount of grain in her dish before she hops onto the stand. If she is a slow eater, you may place her entire portion of grain in the feed pan at this time, but most goats will devour the feed very quickly. I feed approximately 1 to 1 ½ cups of grain at each milking.
2. Brush the underside of the goat and belly to remove loose hair and dirt. In warmer months, you may consider hand clipping the longer hairs, or use an electric clipper to shave the udder area.
3. Sweep hair and dirt off the stanchion.
4. Wash your hands.
5. Wet a paper towel with the udder wash. Wash the udder and teats and discard the towel. You can use hand towels for this purpose, which will require you to machine wash the towels. The most important thing is to prevent contamination from udder to udder by not using the same towel on more than one goat. Do not re-dip the used towel into the udder wash. Always use a new towel.
6. Dry the udder completely.

7. Squirt a stream of milk from each teat into the strip cup. This removes any milk that has accumulated at the base of the teat between milkings. It flushes the teat out, and gives a concentration of milk to examine.
8. Evaluate the milk in the strip cup. Look for clumps, strings, flakes, blood or pink milk. Is the milk a normal consistency, not too thin or thick? If the milk does appear to have issues, test for Mastitis (see pages 102 and 220). Always discard milk collected in the strip cup.

How to Milk by Hand

1. Place the stainless steel bucket on the milk stand or stanchion, slightly in front of the udder. If you are milking a small goat, or a goat that produces a small volume, you may hand hold a container up to the udder while milking with the other hand.
2. Sit down or stand up to milk, on the left or right side. Your comfort will be important in this process. The more comfortable and calm you are, the smoother the milking process goes.
3. Grasp a teat where it meets the udder. You will make a circle around the top of the teat with your thumb and your index figure (as if you are signaling OK!). Do not include the udder in this grasp.
4. Squeeze your thumb and index finger together. This step closes the teat from the udder, traps and forces the milk into the teat. When milking a cow, the teats are pulled. When milking a goat, it is a squeezing action. Pulling can damage the teats and udder of a goat.

5. Squeeze your middle finger to force the milk of out the teat. Picture how you hold a teat by holding our hand in the air, making the OK signal with your thumb and index finger, but hold it in a horizontal position. Bring your middle finger in so that it rests and crosses just below where the thumb and index finger meet. Squeeze the teat with your middle finger to force the milk out. Practice makes perfect in this step. Sometimes the squirt does not land in the right direction. Think of this as a two-step process, squeeze the upper teat and release slightly, while immediately squeezing the middle and release. Do not think too hard about what you are doing, except to guide the milk. You will pick up the rhythm.

6. You can alternate from teat to teat, or milk from both teats at the same time, whatever works best for you. Remember, squeeze, squirt, squeeze squirt. Do not pull. You will make mistakes. Do not wear your best clothing...milk may fly! I have milked down a sleeve, across the room, and practically into the mouth of a waiting cat.

7. Continue milking until the milk volume decreases. Urge the doe to drop more milk by massaging her udder, or by bumping it. To bump, lightly use your knuckles (as if you are knocking on a door), or use the palm of your hand. Do not use enough punch to cause pain or bruising.

8. The doe may drop another 4 – 10 ounces of milk after she is bumped.

9. Once the teats begin to feel flaccid, stop milking. To avoid Mastitis, do not completely strip the remaining milk.

10. Cover and move the milk bucket to the side, away from the milk stand. Dip or spray the teats and release the goat. Call in the next goat, and repeat the pre-milking and hand milking steps.

Milking by Machine

Machine milking follows most of the same steps as hand milking. Follow the manufacturer's instructions for machine milking and cleaning. Ensure the pulse and vacuum is set for a goat. Incorrect settings can damage a goat's udder.

Additional Milking Do's and Don't's

- Never pull on a teat while milking, especially from the end of the teat. This is uncomfortable and can be damaging.
- Some goats eat quickly while on the milk stand, and they have the remarkable ability to stop their milk flow as soon as they finish eating. In these situations, and it will happen over and over again with the same goat, give her small amounts of feed, a little at a time, instead of filling the pan with the entire ration at once.
- You can milk while standing up, depending upon the height of your stanchion, or you can milk sitting down. You can also milk on either the left or the right. You will find, however, that individual goats prefer one side or the other. Some goats do not like you to reach under the belly. She may milk more peacefully if you milk from behind. Go with the flow, with the milking position that works.

- Occasionally, some dairy goats overly engorge with milk, to the point the milk stream begins before milking begins. Imagine, milk squirting across the milk stand when she climbs up. This describes a goat that must be milked twice a day, and consider milking her at 12 hour intervals. This type of engorgement normally does not last long. It will likely hit at the 3rd or 4th week, and level off in the 2nd month.
- To keep a semblance of order in your barn, always milk the goats in the same order.

Handling Goat Milk

As soon as the milk is collected, it must be filtered, and chilled. The quicker the milk cools to optimal storage temperature, 40 degrees within 45 minutes, the less likely it will spoil.

If you do not have time to follow all of the Milk Handling Steps immediately after you finish milking, place the tote or bucket inside a larger container that is filled with ice or ice water. This helps the milk to cool down while you are finishing chores. Note: when the ice cubes have melted it is a good indication that the milk has cooled properly. You may also place the milk in a freezer for quick cool down.

Milk Handing Steps

1. Wash your hands or wear clean gloves.
2. Ensure the strainer, containers, and the lids are sanitary.
3. Place the milk filter in the strainer.
4. Place the strainer over the opening of the milk storage container. This can be a large glass jar, a stainless steel tote, or several smaller glass jars. The smaller the storage container, the quicker the milk will cool down.

5. Fill the storage container by pouring the milk from the bucket or tote through the strainer.
6. Place airtight lids on the milk storage containers and store the milk in the refrigerator or freezer. To avoid contamination, discard plastic containers after one use. Plastic etches which makes sanitation difficult. You may, however use sanitized plastic lids on glass containers in the refrigerator.

These instructions are for the basic non-commercial dairy farm. On a commercial farm, specific state requirements must be followed.

Heat Treating and Pasteurization

I am a fan of drinking raw milk produced on my own farm. I know what goes into my goats, and I know the milk produced by my healthy goats is both wholesome and safe. Goat milk is sweet and tastes untainted when handled properly.

Many states do not allow raw milk sales, or pasteurized milk sales without a license. Heed to these regulations, you do not want to lose your rights to your dairy farm.

To prevent disease, either heat treat the colostrum or pasteurize the goat milk.

Heat Treat Process:

1. Filter the colostrum and gently bring it to 135°F (57.2°C).
2. Hold colostrum at no greater than 135°F (57.2°C) for 60 minutes to kill pathogens and bacteria.
3. Cool the colostrum by placing the container in an ice bath or freezer. Refrigerate or freeze as soon as possible.

The best heat treat method for colostrum is to use a double boiler, or pour the colostrum into a zip lock bag and place it in a slow cooker filled with water brought to 135 degrees Fahrenheit.

Colostrum thickens and solidifies at higher temperatures, and it may lose its antibody properties.

The Pasteurization Process

There are two common temperatures and holding times for milk pasteurization: High Temperature Short Time (HTST), and the slower lower temperature method, which is United States FDA (Federal Drug Administration) legal.

HTST Pasteurization Method:

1. Filter the milk and bring it to 161°F (72°C).
2. Hold the milk at this temperature for 15 seconds
3. Cool immediately, bringing the milk down to 40 degrees as quickly as possible.

Slower Pasteurization Method:
1. Filter the milk and bring it to 145°F (63°C).
2. Hold the milk at this temperature for 30 minutes.
3. Cool immediately, bringing the milk down to 40°F as quickly as possible.

If you plan to pasteurize milk, you can do so in your kitchen, but I recommend purchasing a store bought pasteurizer. Check your local farm stores and classifieds for used machines. Some machines have a single setting for pasteurization; others have a second setting for heat treating.

Milk Storage

Properly handled milk will keep in the refrigerator for several weeks.

Using a permanent marker, write the collection date on the colostrum or milk container before placing it in the freezer. Keep frozen milk and colostrum for up to one year. Freezing may change the texture slightly, but the taste will remain the same.

After Milking, the Cleaning Process

Farming is normally dusty work. Sweep the stanchions and milking area after and prior to each use.

Always rinse milking utensils and buckets in cool or cold water before washing with hot water. This rinses the majority of the milk proteins from the surface, which will harden and stick if subjected to hot water.

If you are a non-commercial dairy farm, sanitize equipment in a diluted chlorine bleach formula or use a commercial sanitizing rinse agent. If you use equipment that is small enough to fit in a domestic sized dishwasher, wash the pre-rinsed equipment in a dishwasher set on high heat. Most dishwasher detergents contain a form of bleach that sanitizes as it washes. Avoid the use of highly fragranced detergents as they may taint the taste and smell of milk.

Use ¼ to ½ cup of bleach per 4 gallons of hot water as a general sanitizing agent on the farm. A commercial blend, however, will help to retain the finish on milking equipment.

Unless you use the dry heat cycle on a dishwasher, always allow milking equipment to air dry. Place equipment on a wire shelf, or hang to dry, with the open side of buckets and containers facing down.

Mastitis and the Dairy Doe
Mastitis affects dams that are in milk, regardless of hand or machine milking, or nursing (see Mastitis, page 102).

Mastitis Prevention – Things to Remember When Milking:
- Examine the udder and milk each time a goat is milked. Look for lumps and redness of the udder, milk that is "off" in texture, or milk that contains blood.
- Blood in the milk is not necessarily an indication of Mastitis. It may be an indication of rough handling or a broken capillary.
- Always wash and dry the udder before milking.

- When machine milking, remove the suction cups (inflations) promptly, as soon as milking is complete.
- Use a teat dip following milking to prevent bacteria from entering the orifices.
- Do not use the same towels or wipes from goat to goat, or re-dip used towels into the udder cleaning solution.
- Goats infected with Mastitis are susceptible to repeat infections. In the one and only case that I have experienced, I treated successfully, and it never returned. Prompt treatment, cleanliness, and good nutrition are the success factors.
- If you suspect that a goat has Mastitis, milk what she produces into a separate collection container and discard the milk until you are sure (through testing) that she is no longer infected. Do not milk her with a machine, which can further irritate the teats and udder, and can contaminate other goats.
- Consider a loafing area, with fresh hay to munch on for the dairy does. When you are finished milking, this gives the goat incentive to stand up for a while, which prevents bacteria from entering the teats.

Udder Wash Formulas

- Dish soap diluted with warm water, up to a tablespoon of liquid dish soap in a quart of warm water. Can be drying to the teats and udder, but will be mild compared to a bleach solution. May not remove all bacteria.
- One part chlorine bleach to ten parts water. May be drying to the udder and teats.
- Unscented baby wipes that do not contain alcohol. May not remove all bacteria – wash the udder and teats well.
- Concentrated pre-made udder wash (some contain Cholorhexadine). Dilute per manufacturer's instructions. Will clean and eliminate bacteria.
- Apple cider vinegar diluted with water, with a few drops of dish soap added. Up to a tablespoon of apple cider vinegar in a quart of warm water.

Use thick paper towels that resemble soft cloth, washable terry or cotton towels, or blue shop type of towels as udder wipes.

Containers are available in farm stores, and online, that contain large pumps, which dispense liquids in increments of 1 ounce or more. Consider using these for diluted udder wash formulas.

Teat Dip Formulas

- Commercial spray that disinfects and helps the teat orifice close
- ½ cup of hydrogen peroxide, and 1 tablespoon of aloe juice, diluted in ½ gallon of warm water
- ¼ teaspoon of chlorine bleach, and 1 drop of mild dish detergent, diluted in 1 quart of warm water

Chlorine bleach mixed with soap, and other chemicals, can be dangerous, toxic, with fumes that burn the eyes and breathing passages, mix with extreme caution.

Udder Balm Formulas

- Ready-made udder balm
- Organic coconut oil
- Unscented lotion, or for extra soothing and massaging, add 5 drops of peppermint essential oil to 16 ounces of lotion
- 10 drops of peppermint essential oil, and 5 drops of tea tree oil, mixed with 1 cup of coconut oil

Udder balms containing fragrance or essential oils can affect the taste of the milk.

Practice Milking with a Glove

The glove practice method is a way to learn how to milk without involving a goat.

1. Fill a strong disposable glove with water and tie the opening at the wrist off.
2. All of fingers of the glove have now filled with water, which gives you an idea of how it feels to milk goats with different sized teats.
3. Hold the glove with one hand.
4. Place your other hand up against the udder (the palm of the glove).
5. Where the teat (finger) meets the udder (palm), place your thumb and index finger, as if you are making the "okay sign."
6. Use these fingers to close off the teat from the udder. When you squeeze these fingers, it closes the milk off from the udder and forces the milk into the teat – towards the tip.
7. Tighten your middle finger to force the milk down into the teat, which with a real teat, would force the milk to stream out of the teat.
8. If you can feel water backwashing into the udder, you are holding your fingers incorrectly.

This is really a simple process. Squeeze with two fingers to close the flow off from the udder. Squeeze with the middle finger to force the milk out. Trust me, you will develop rhythm!

Drying off a Dairy Doe

This is one of the most frequently asked questions, how do I dry off my dairy doe?

Stop milking her. It is practically that simple.

Cut the grain out of her diet, or decrease the amount down considerably. If she is on the slender side, allow her to continue visiting the milking station, and feed her a small amount of grain. Keep in mind, once you stop milking, she will put weight back on.

The common concern when drying off a dairy doe is udder engorgement, "Her udder is so large! I feel sorry for her!" The udder will look uncomfortable, but this does not lead to Mastitis.

The sheer act of milking causes the goat to produce more milk. Eventually, when milking has stopped, the engorgement decreases.

Mary L Humphrey

Chapter 14
Ways to Use Goat Milk

What do I do with all of this milk?

Freeze it for future bottle kids. Drink it. Feed it to the pigs. Use it for cheese, butter, or soap!

Cheese

Ask around, how many people do not like artisan goat milk cheese? Goat cheese varieties are nearly endless – Chevre', cottage, and feta, are especially wonderful when infused with chives and fresh herbs. Hard cheeses, such as cheddar, are wonderful when made from goat milk. Cheese making equipment, including presses, is readily available online through dairy and goat supply businesses.

Simple Microwave Mozzarella Cheese Recipe

1 gallon goat milk
1.5 to 2 tsp. citric acid
¼ tsp. liquid rennet (already diluted with water per supplier instructions)
½ cup cool water (not chlorinated)
Kosher salt to taste (not iodized)

Equipment needed: thick rubber gloves, stainless steel or enamel cooking pot, glass bowl, spoons (for stirring), large knife for cutting

Dissolve the citric acid in ¼ cup water

Pour the milk into a stainless steel or enamel cooking pot. Add the citric acid solution. Stir well. Slowly heat the milk to 90°F.

The milk will begin to curdle.

At 90°F, add ¼ tsp. liquid rennet solution. Stir gently, in a top to bottom motion for approximately 1 minute.

Remove the pot from the heat source.

Allow the milk to set for 5 minutes. This allows the curd to form.

With a long knife, cut the curd in a checkered or cubed pattern, approximately 1" in size.

Scoop the curds into a glass bowl.

Press the curd with your clean or gloved hands. The goal is to release and pour off as much liquid (whey) as possible.

Microwave the curd on high for 1 minute.

Caution, the cheese and bowl will be hot! Wear thick rubber gloves during this process.

Remove the bowl from the microwave and quickly knead cheese with a spoon, or your gloved hands. Drain off the whey again.

Microwave 2 more times, 35 seconds each. Repeat the kneading, handling and draining each round.

Knead quickly, as you would bread dough. You will begin to see the cheese firm up, and it will become shiny. Add salt at this time, to taste.

The cheese should now be pliable and stretchable.

Form the cheese into a ball and drop it into ice water to cool, or roll it out and wrap it in wax paper. I prefer the wax paper as it does not subject the cheese to water. Place the wrapped cheese in the refrigerator to cool.

The cheese is ready to eat once cooled.

Tightly wrapped and refrigerated cheese will stay fresh for many days.

Note:

- Whole cow milk can be used for this recipe, but the taste will differ from goat cheese
- Lipase powder may be added (to give a stronger Italian taste)
- Do not use aluminum pans when making cheese
- Do not use iodized salt, it causes the cheese to take on a green hue
- Cheese making equipment must be extremely sanitary prior to use. Any amount of dirt or contaminant will ruin an entire batch of cheese. It will not form into a solid cheese, or it will spoil rapidly.

Butter

People say, "You can't make butter from goat milk." I have. What this requires is rich butterfat milk, milk that separates, with a film of cream floating on top 24-48 hours after it has thoroughly chilled in the refrigerator.

I make fresh butter by scraping off the goat milk cream and placing it in a sterilized glass jar with tight fitting lid. I shake, roll and move the jar until the solids (butter) separate. I then drain off the liquid (buttermilk). Add salt if preferred, preferably Kosher, as it does not affect the color or flavor of the butter. Butter can also made with a mechanical separator. Do not invest in this equipment unless you have a large volume of milk that is high in butterfat.

Goat Milk Soap

Goat milk soap is another way to utilize extra milk. Cheese requires a large amount of milk, up to 1 ½ gallons for an average family-sized recipe. Soap, however, requires much less milk for a batch that makes 9-18 bars.

Goat milk soap is soothing, it is gentle, and it does not promote dry skin.

Goat milk soap is my specialty, see page (i) for a list of my soap and lotion making books.

Chapter 15
Earning an Income with Goats

Making a profit from goat ownership is not easy, but it is obtainable with a plan in place and determination. In other words, run your farm as a business and you will experience results.

Before you pin-point how much income you plan to obtain through your goat farm, determine your goals. When income exceeds expenses, your farm is no longer considered a hobby, it becomes a profit-making business. Ask yourself, do you want to break even, meaning, do you want the goats to pay for themselves? Do you expect a small or large profit above your farm expenses?

List your farm expenses that are directly related to raising goats, such as, feed, hay, water, electric, maintenance of fencing, housing, barn, medications or other supplies, milking equipment.

List the ways you can make money with your herd, such as, milk or cheese, kids, farm visits, weed eating or brush clearing (involves taking your herd to the location), goat milk soap, classes, speeches, seminars, books, and semen sales.

Mock up a business plan, a platform. How are you going to make money? How often are you going to bring the income in? What do you need to do to bring in the income (marketing, advertising)? How much revenue will you be bringing in?

Compare the expenses to the projected income, month by month. Will you obtain the profit that you expect?

There are also ways to keep expenses down, which is very important during times of poor economy:

- Set a maximum number of goats for your herd – sell every goat above that number, and sell as soon as possible after new kids are weaned
- Research feed costs – determine the best nutrition at the lowest cost. Feeding poor feed, however, will result in lost revenues, and sick goats. Expect variances in feed costs, especially increases in grain and hay prices. Today's inclement weather can affect the price of future feed. Once the price increases, it generally does not decrease.
- Learn as much as you can about veterinary care. A sustainable farm requires hands-on animal care. Veterinary assistance can be expensive and generally is requested for emergencies only.
- Plan ahead – a successful entrepreneur has policies, goals, and plans in place (breeding, birthing, vaccinations and health maintenance, feeding, nursing, selling – including policies and guidelines for pricing, advertising and return policies)

- Take small steps with growth. Goats and businesses (like a soap business) can be addictive. Meaning, we follow our passions but forget that we are trying to make a profit. Taking small steps ensures you do not spend your money or time in too many places, too quickly.

The most important thing to remember is, treat your farm as a business if you desire a profit. You can care for your animals, love them beyond means, but still make a profit, or break even, if you have a plan in place and you stick to it.

You may also consider offshoot businesses such as feed distributor, animal care consultant or caretaker, additional farm produce (eggs, hay, fruits or vegetables, herbs, fiber, or spun yarn).

Mary L Humphrey

About the Author

Mary is an author, blogger, life coach, and has had a life-long interest in animals.

She raised a herd of dairy and Boer goats for nearly a decade, which included a successful goat milk soap and lotion business, and she experienced personal satisfaction in teaching others how to raise and milk goats from her own barn. She often received calls to assist other goat owners with health care and with the birthing process.

Mary is an advocate for humane animal care and dairy farms around the world.

Mary began her love of goats from a young age when she would linger in the petting areas of zoos and farms with Pygmy goats. Later in life, while she and her husband were purchasing hay for their horses, a herd of white animals with dark brown heads bounded to the nearby fence. What were those excited, happy, and healthy creatures? They were Boer goats, one of the first herds that had arrived in the heartland of the United States. Mary's deep interest in goat-raising began that day.

Mary researched books, magazines, and online articles. When she visited the local veterinarian, she stood side by side to learn all that she could. She furiously studied goats for several years, which led her to today's passion, to write this book and help others to start out with goats, and even to give a new perspective to those that have raised goats for many years.

From Mary, "Goats are wonderful creatures. They have personality. They have humor, and they can be great companions. Goat milk is remarkably full of perfect nutrition, and the products that can be produced from goat milk (yogurt, cheese, and soap) are rich and beautiful."

"I hope you enjoy this book as much as I have in writing and sharing it with you!" Mary

The Editing Team

Sharon Parker is a born and raised in the country person. She resides on a 20-goat and horse farm in Lexington, North Carolina. In her words, "Loving animals is in my blood."

An addition to Sharon's long-lived passion for animals, for riding and caring for horses, came in May 2001 as a single black and white Nigerian Pygmy mix goat named Jewel. As a child, Sharon dived into her daily home-school work with gusto knowing that she would spring out the door at completion to take care of Jewel. She learned to love and care for goats from this initial experience. The goat farm grew to include Nubians, taking care of kid goats and milking, and Sharon now enjoys helping others care for their goats as well. She says, "I am addicted."

Today, Sharon is busy developing a goat milk soap business, and can be reached at goatsrule2003@yahoo.com.

***

Starla Ledbetter resides on a small farm in Northern California where she raises LaMancha dairy goats and makes goat's milk soap and other bath and body products. Her adventure with dairy goats began in 2005 as a 4-H project for her daughter and then turned into a passion of her own.

Starla can be reached at starla@rooranch.net, www.rooranch.net, or join her on Facebook at www.facebook.com/rooranch.

***

Debbie Richards is a professional soapmaker, business owner, and freelance writer and editor. Her professional career has spanned more than two decades and includes positions as a software trainer, help desk manager, technical writer/editor, research and development lab technician and quality assurance manufacturing manager. Debbie's eclectic interests include cooking from scratch, mentoring others on healthy life-style living, organic gardening, and playing guitar. She recently realized her lifelong dream and moved to a farm in Southwest Michigan.

Debbie's website is www.bigfatsoap.com and you can read her blog at www.othersideofdenial.com. Her email is deb.richards@yahoo.com.

Appendix 1
Goat Care Resources and Suppliers

Businesses that carry goat care and farm supplies:

Hoegger Supply Company
www.hoeggerfarmyard.com
800-221-4628
Online shopping
Goat management, milking products, cheese making

Jeffers
www.jefferspet.com
800-533-3377
Online shopping
Goat management

Caprine Supply
www.caprinesupply.com
800-646-7736
Online shopping
Goat management, milking products, cheese making

Hamby Dairy Supply
www.hambydairysupply.com
816-449-1314
Online shopping
Goat management, milking products, cheese making

PBS Animal Health
www.pbsanimalhealth.com
800-321-0235
Local stores and online shopping
Goat management

TSC (Tractor Supply Company)
www.tractorsupply.com
Local store finder online
Goat management

Orscheln Farm and Home
www.orschelnfarmhome.com
Local store finder online
Goat management

Buchheit
www.buchheitonline.com
Local store finder online
Goat management

Sydell
www.sydell.com
800-842-1369
Online shopping
Goat management equipment

Lakeland Group
www.sheepandgoatequipment.com
855-886-8382
Online shopping
Goat management equipment

Appendix 2
Goat Associations

American Dairy Goat Association (ADGA)
www.adga.org

American Goat Society (AGS)
www.americangoatsociety.com

The American Meat Goat Association
www.meatgoats.com

American Boer Goat Association (ABGA)
www.abga.org

United States Boer Goat Association
www.usbga.org

Appendix 3
Recommended Resources

Purdue University
Department of Animal Sciences
www.ansc.purdue.edu/goat
Goat care and disease

Cornell University
www.albany.cce.cornell/resources/kidding
Photographs of birthing and goat care

Purdue University
Department of Animal Sciences
www.extension.purdue.edu/exmedia/id/id-321-w.pdf
Hoof trimming

Dairy Goat Journal
www.dairygoatjournal.com/81-2/staff_report/
Illustration of birthing positions. I recommend
subscribing to this magazine, especially if you plan to
own or show a dairy goat.

International Kiko Goat Association
http://www.theikga.org/how_to_give_injection.html
Illustration of injection sites

Mary L Humphrey

Glossary
Common Goat Related Terms and Words

Abortion – Offspring born before they are able to live outside of the womb.

Anemia (Anaemia), Anemic – A condition in which the blood does not have enough healthy red blood cells.

Anorexia – A loss of interest in eating.

Billy – Slang word for adult male goat.

Browse - Leaves and plant material produced from trees and tall plants.

Buck - Intact (not castrated) male goat.

Buckling - Intact (not castrated) young male goat.

Caprine - Scientific name for the goat species.

Cashmere – Fiber from the coat of a Cashmere goat.

Castrate – Removal of the testicles of a male goat.

Chevron – Goat meat.

Cud – Feeds regurgitated from the rumen that the goat chews and swallows.

Cull – A management practice to remove or sell goats from a herd.

Dam – Mother goat.

Disbud – Process of removing horn buds from young kids, which prevents horns from growing.

Doe – Female goat.

Doeling - Young female goat.

Dorsal – The back side, pertaining to the area of the back bone.

Dry Off - The process of taking a doe "out of milk," no longer milking her to stop her from producing milk.

Fiber - Long hair produced by a goat, called mohair or cashmere.

Freshen – To give birth and begin producing milk again.

Gestation – The period of time between conception and birth.

Goat Berries – Common name for goat feces.

Graze - The act of eating pasture.

Herbivore – An animal that gets it energy solely from eating plants.

Hermaphrodite – An animal that has reproductive organs associated with both male and female sexes.

In Milk – A doe with an udder that is producing milk.

Incubation Period – The amount of time it takes for symptoms to appear after an animal is exposed to a disease or pathogenic organism. The animal may or may not be contagious during the incubation period.

Intact - Bucks that have not had their testicles removed.

Kid - Young male or female goat offspring.

Kidding – Birthing, or "to kid," will give birth.

LGD – Livestock guardian dog.

Muzzle - Lower jaw area.

Nanny – Slang word for adult female goat.

Necropsy – Term for animal autopsy.

Offspring – Kids, the goats that a doe or buck conceived.

Parasites – Internal or external organism that lives on a host.

Pathogen – Infectious or biological agent that causes disease or illness to its host.

Parturition - The act of giving birth.

Polled – A goat that lacks horns due to genetics.

Quarantine – Separation of animals from a herd to prevent the spread of infectious disease.

Rumen – The first of four stomachs in a ruminant's digestive system.

Ruminant – The term for any animal (ex: goat, cow) with four stomachs, the first of which is the rumen.

Sire – Specific buck that fathered offspring.

Teat – The nipples on a goat's udder from which a kid nurses, or from which a goat is milked.

Tethering – Tying an animal to a stationary item by a rope or chain.

Udder – The mammary gland of female goats (cows, sheep and horses). The udder of a goat contains two compartments and two teats.

Wean - The process of removing kids from dependency of milk, accustoming them to pasture and solid feeds.

Wether - Male goat that has had their testicles removed, which eliminates their ability to breed.

Annie's all about Goats

Made in the USA
San Bernardino, CA
15 July 2016